Earth and Environmental Sciences Library

The *Earth and Environmental Sciences Library* (EESL) is a multidisciplinary book series that showcases innovative approaches and in-depth reviews, aimed at enhancing the role of the Earth and Environmental Sciences communities. The series also offers valuable guidance for stakeholders, decision-makers, policymakers, international organizations, and NGOs.

Topics of interest include oceanography, the marine environment, atmospheric sciences, hydrology, soil sciences, geophysics, geology, agriculture, environmental pollution, remote sensing, climate change, water resources, and natural resource management.

To further advance these critical areas, the *Earth and Environmental Sciences Library* invites contributions from the Earth and Environmental Sciences communities in the form of edited volumes, monographs, textbooks, conference proceedings, and more. The series fosters knowledge exchange and collaboration across disciplines, encouraging comprehensive discussions and solutions to pressing environmental challenges.

Tatiana Chaplina
Editor

Coastal Geosystems of Cold Seas Under Conditions of Climatic and Anthropogenic Changes

Springer

Editor
Tatiana Chaplina
Zagromonte SL, Carretera
Alicante, Spain

ISSN 2730-6674 ISSN 2730-6682 (electronic)
Earth and Environmental Sciences Library
ISBN 978-3-032-12246-9 ISBN 978-3-032-12247-6 (eBook)
https://doi.org/10.1007/978-3-032-12247-6

This Springer imprint is published by the registered company Springer Nature Switzerland AG
The registered company address is: Gewerbestrasse 11, 6330 Cham, Switzerland

Preface

This book is devoted to the most important and urgent problems that arise during the development of the coasts of the cold seas. The most important result of the work is the creation of an engineering-oriented concept of morpholithodynamics of the shores of the Arctic and subarctic seas and the forecast of coastal development in conditions of rapid natural and anthropogenic changes. The established patterns of development of the coastal zone in the coordinate system of the concept of contact zones of different types of seas and seas and ocean open up new opportunities for improving the theory and methodology of cognition of the natural environment and the relief of the coastal zone as the basis for the life of society and ensuring sustainable development in these regions. In addition to the theoretical significance, the established patterns have a pronounced applied aspect—solving infrastructure safety problems, coastal protection and marine logistics, risk assessment and planning of economic activities on the coast, and creation of natural and economic accounting systems. Using the example of the Far Eastern region of Russia, the issues of anthropogenic transformation of coastal morpholithosystems are considered. The results of theoretical constructions and modeling of morpholithodynamic processes are used to determine the quantitative parameters of the carbon cycle of biomorpholithosystems of the coastal zone and adjacent marine areas, in conditions of accelerating evolutionary changes and increasing anthropogenic load. This makes it possible to switch to controlled climatic activities for the absorption of carbon dioxide from the atmosphere, accumulation, transformation, and long-term burial of autochthonous and allochthonous organic carbon using nature-like technologies on created coastal marine wetlands (carbon farms). The book is addressed to specialists in the development and implementation of projects related to investment, construction, natural and climatic activities, modernization of housing, communal, industrial, transport, social, energy, and other infrastructure in coastal areas in high latitudes. It can be useful for students and postgraduates studying in the specialties of geomorphological engineering, hydrotechnical, and climatic directions.

Alicante, Spain

Tatiana Chaplina

Contents

Study and General Characterisation of the North Pacific Seashores 1
V. Afanas'ev Viktor, E. I. Ignatov, and A. B. Faustova

Morphology and Dynamics of the Shores of the Contact Zone of Temperate and Subarctic Seas 37
V. Afanas'ev Viktor and E. I. Ignatov

Patterns of Coastal Destruction at the Present Stage 83
V. Afanas'ev Viktor

Beach Formation and Distribution of Its Wave-Absorbing Properties along the Shore 145
V. Afanas'ev Viktor and E. I. Ignatov

Conclusion 181

Study and General Characterisation of the North Pacific Seashores

V. Afanas'ev Viktor, E. I. Ignatov, and A. B. Faustova

Abstract The coastal zone is a vital resource for humanity, and its ecological health is crucial for the sustainable development of coastal regions. While the fundamental processes of coastal dynamics are universal, significant regional variations exist. The shores of Russia's Far Eastern seas, which span arctic to temperate zones, present a unique and urgent case study.

Historically, industrial development began during a period of weak coastal activity and sediment accumulation. However, a widespread increase in erosion in the latter half of the twentieth century, combined with a disregard for natural forces, created significant risks. The recent expansion of the oil and gas industry has intensified the need for better understanding, generating vast amounts of geological data. This data must now be synthesized to advance geomorphological science and inform practical applications. A key challenge is that the behavior of these Subarctic shores often defies traditional theories, as they have rarely been studied as independent systems.

Keywords Geomorphology of the coasts · Far Eastern seas of Russia · Human exploration of the coast · Regional features of coastal morpholithogenesis

The North Pacific is renowned for the major geographical discoveries of its islands and territories; however, scholarly assessment concludes that the study of its shores is inadequate. The initial era of discovery eventually gave way to systematic general geographic and targeted geomorphological investigations of the shores and seabeds of the Far Eastern seas—including the Bering, Okhotsk, and Japan Seas—and the islands of the Aleutian, Kuril, and Japanese arcs (Table 1). This intensification of scientific research was driven by the expansion of economic activity and the

V. Afanas'ev Viktor (✉) · A. B. Faustova
Sakhalin State University, Yuzhno-Sakhalinsk, Russia

E. I. Ignatov
Faculty of Geography, Moscow State University, Moscow, Russia

T. Chaplina (ed.), *Coastal Geosystems of Cold Seas Under Conditions of Climatic and Anthropogenic Changes*, Earth and Environmental Sciences Library, https://doi.org/10.1007/978-3-032-12247-6_1

Table 1 Stages of North Pacific Coastal Studies

Stages of research	Researchers	Results
Expeditions of the USSR IOAS and Lomonosov Moscow State University (MSU) on the shores of the far eastern seas. Study of earthquake and tsunami consequences	Vladimirov A. T. (1959a, 1961, 1959b); Medvedev V. C. (1959, 1961), Ionin A. C. (1959, 1992), Kaplin P. A. (1961, 1957, 1962), Zenkovich V. П. (1967, 1945, 1971a, 1980, 1953, 1971b), Budanov V. I. (1961), Gershanovich, D. E. (1957), Zabelina, E. K. (1961), Leontiev, O. K. (1975, 1967, 1965), Kanaev V. Ф. (1959), Pavlidis Y. A. (1966), Nevessky E. N. (1967) Miller D. J. (1960), Reimnitz E. (1965), Pararas-Carayannis G. (1967), Wilson B. W. (1966)	Extensive information on the structure and development of the shores of the North Pacific has been obtained and systematised. A complete picture of the structure and development of the Far East coasts was created. Maps of the coasts were compiled for the physical-geographical atlas of the world (1964). The values of seismic dislocations in the coastal zone and the morpholithodynamic consequences of tsunamis were determined
Case studies: Far Eastern State University (FESU), MSU, Siberian and Far Eastern Departments of RAS, MinGeo, etc.	Archikov E. I. (1982, 1979, 1985), A. P. Kulakov (1966, 1972, 1973), L. B. Hershberg. (1982), Korotkiy A. M. (1982, 1990), Kravtsov V. D. (1980, 1979). Weinbergs I. G. (1975, 1990), Grabkov V. K. (1982, 1969), Gurieva Z. I. (1966), Brovko P. Ф. (1985), Bobykina V. П. (2000, 1981), Barkov L. K. (1985), Denisov, S. V. (1967, 1966), Polunin, G. B. (1977, 1986), Remenets, A. I. (1975, 1981), Kononov Y. I. (1975a, 1969, 1992, 1975b). Ignatov E. I. (2005, 2004, 1976, 1979, 2000), Hayes, M. O. (1977), Molnia, B. F. (1977), Jordan J. W. (1999), Mason, O. K. (1997), Lander, J. F.(1995), Boothroyd, J. C. (1975)	An evolutionary scheme of coastal development in the Neopleistocene-Holocene was created, seismoacoustic and lithological studies of the underwater coastal slope were carried out, and the first quantitative data on the intensity and activity of modern coastal processes were obtained. Based on mineralogical and granulometric analyses and the LMP method, lithodynamic maps of the coastal areas of Sakhalin and Primorye were compiled
Study of the shores of the North Pacific at the present stage	Gorin S. Л. (2015, 2012), Korotkiy A. M. (2011), Skrylnik G. P. (1979), P. McLaren (2019, 2015), Sunamura T. (2015), Mason, O. K. (2012), Jones, B. M (2009), Gorokhovich, Y. (2011), Gibbs, A. E. (2015a, 2015b, 2017), Uda T. (2010), Afanas'ev, V. B. (2020), Leontiev, I. O.(2015)	The climatic component was updated and the subarctic type of coastal morpholithogenesis was identified. Remote sensing methods and high-precision photogrammetry allowed to obtain quantitative characteristics of the dynamics of accumulative forms, scouring of marine terraces and destruction of rocky shores

development prospects of coastal territories, whose rich natural resources—such as the offshore oil and gas deposits near Sakhalin Island—stimulated mineral exploration.

From the first permanent settlements and throughout the period of coastal development, ideas about the dynamics and development of the coasts of the seas and oceans were formed. But only in the XIX century, according to V. V. Longinov, there are "...the first attempts to create a unified concept of coastal zone dynamics" (1959).

In the XX century the fundamental works on coasts and development of coastal processes by V. P. Zenkovich (1945; 1962), O. K. Leontiev (1955), V. V. Longinov (1963), and then appear the works of V. I. Budanov (1964), P. A. Kaplin (1973), N. I. Aibulatov (1965), Y. S. Dolotov (1982), G. A. Safyanov (1976; Safyanov 1973) and others.

In recent years, R. D. Kosyan, I. O. Leontiev, L. A. Zhindarev, N. V. Esin, E. I. Ignatov, S. A. Lukyanova, E. N. Badyukova, T. A. Yanina, and others have made a great contribution to the development of the theory of marine shore formation and modelling of morpholithodynamic processes. However, despite numerous attempts to classify shores by genetic or dynamic types, it has not yet been possible to create a conceptual model of coastal processes development, trend and stages of seashore development in high latitudes, especially in the conditions of the contact zone of transition from subarctic to temperate-cold seas.

1 General Physiographic Characterisation, Zoning and Typification of the Shores of the North Pacific Ocean

As G.P. Skrylnik notes, "the Far East is located in a steadily active transition-contact zone of the two greatest structures of the Earth, in the course of interaction exerting powerful system-forming influences within the temperate, subarctic and arctic natural-climatic belts" (Skrylnik 2019). At the same time, the natural tendencies in the development of physiographic provinces are continentalisation of the natural environment against the background of regional cooling (Nikolskaya 1975; Nikolskaya 1974; Skrylnik 2004).

The boundaries of geosystems (GS) are relatively mobile zones of mutual influences and interactions of neighbouring GS—contact geographical structures (Likhacheva 2010; Baklanov 2014). Most geomorphological boundaries can be regarded as a type of contact zones identified in physical geography, and the vast majority of earth processes and phenomena are confined to them (Likhacheva 2006, 2007).

The geographical position of the Far East on the eastern edge of Eurasia on the border with the Pacific Ocean determines the presence here of the monsoon circulation, which can be traced to about 70°N (Dzerdzeyevksii 1975; Tunegolovets 2006). The peculiarities of the monsoon circulation are determined by the interaction of the three main monsoon regions (Lisogurskii 1980). The polar monsoon is developed in a narrow band along 70°N. From 45° to 65°N is the area of the temperate monsoon. The area of stable subtropical monsoons is located south of the 45th parallel.

The interaction between land and ocean is carried out by the interaction of three climatic centres of atmospheric action (the Siberian and North Pacific anticyclones and the Aleutian depression), as well as seasonal centres—the Okhotomorsk anticyclone and the Amur depression (Tunegolovets 2006).

The continental influence is manifested mainly in the form of northwestern and northern continental air flows (winter monsoon), which strongly reduce the warming effect of the Pacific Ocean on the coast, while the ocean influence is manifested by southeastern and southern sea air flows (summer monsoon).

Other major morphoclimatic factors on the coast are latitudinal and weather differentiation of solar energy radiation inflow, the influence of climatic fronts (Arctic front and temperate front) determining intensive cyclonic activity, tropical cyclones, relief and morphostructural plan of the coasts.

The relief of the Far East is shaped by a complex and often contradictory interaction of continental and oceanic landforming influences, which are dictated by its modern zonal and provincial characteristics. Based on this analysis, three categories of territories have been identified, each characterized by a distinct trajectory of relief development in combination with specific air masses. The morphostructural nature of the region is expressed primarily in the relative youth of its relief, whereas its morphosculptural character is evident in the climomorphogenetic diversity of its geosystems (Nikolskaya 1974; Skrylnik 2006).

The interference of heat and moisture fluxes from the Atlantic and Arctic Oceans is a primary factor in the development of Europe's longitudinal physiographic sectors and the latitudinal zonality of its vegetation cover (Alekseev 2004; Dyakonov 2017).

The morphotectonic structure of the Northern Pacific's continental margins and its development since the Neogene period govern the region's modern coastal processes, including its climomorphogenesis and the latitudinal and longitudinal patterns of its morpholithogenesis.

It should be noted the northeastern orientation of the main morphostructural elements that determine the main features of land-sea interaction and the relative isolation of the Far Eastern seas from the Pacific Ocean as a result of the development of the Kuril and Aleutian island-arc systems (Ermakov 2016; Bark 1970; Kolomyts 2016; Krasny 1965; Hanchuk 2009). This has resulted in a significant counterclockwise shift in the boundaries of climatic zones along the coast. The Sea of Okhotsk is thus categorised as a subarctic sea (Matul 2009; Ustinova 2002; Hen 1997). Accordingly, the coastline of the Sea of Okhotsk has features of Arctic and sub-Arctic coasts. Summer geomorphological processes are influenced by the conditions of the preceding winter. In contrast, the North American coasts of the temperate zone are influenced by the subtropical zone, as coastal morpholithogenesis is influenced by warm ocean currents throughout the year.

The morphoclimatic position (geospatial parameters and associated differences in morpholithodynamic parameters) of Sakhalin Island and the island-arc systems of the North Pacific allowed us to consider them as contact zones of seas of different types, as well as seas and oceans (Fig. 1).

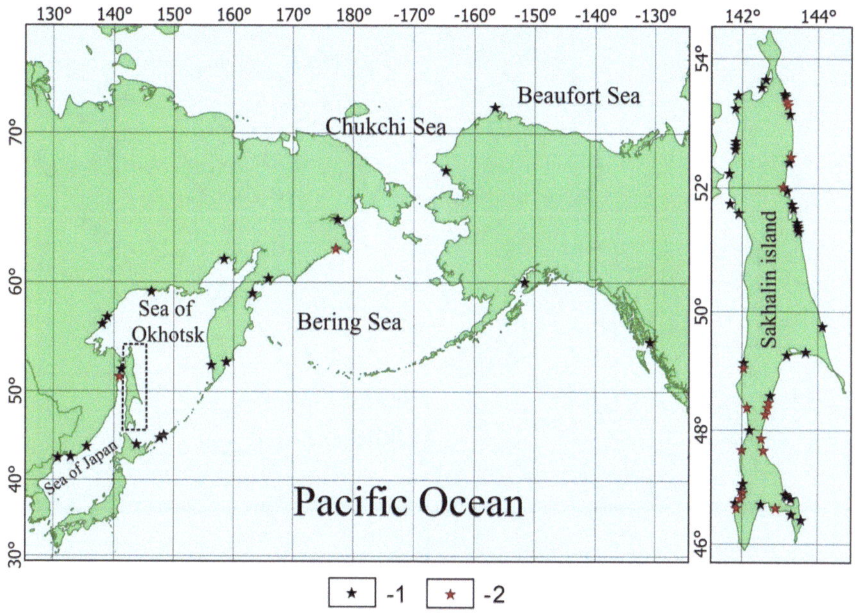

Fig. 1 Overview map. 1—areas of detailed geomorphological studies, 2—areas of engineering-geomorphological substantiation of bank protection

The physical processes shaping coastal zones—such as sediment movement, wave energy, and water temperature—also define the zone's climate-influenced morphological features. A key example is Sakhalin Island, where its shores on the moderately cold Sea of Japan and the subarctic Sea of Okhotsk are separated by just 30–140 km. This is in stark contrast to large islands in archipelagos, where their opposing ocean and sea coasts are often only a few kilometers apart.

Thus, the wind-wave fields and their interaction with the quicksand waves in the contact zones have a very complex character. The temperature regime of the coasts also depends to a large extent on the geospatial parameters of the contact zones (Fig. 2).

The specifics of coastal morpholithogenesis in the seas of subarctic and temperate zones are presented in Table 2.

2 Morphoclimatic Characterisation of Sakhalin Island as a Contact Zone of Temperate and Subarctic Seas

The shores of the North Pacific, whose water circulation is part of the general cyclonic circulation of the Subarctic Pacific, are located within several physiographic zones (Arctic, Subarctic and temperate) (Fig. 3).

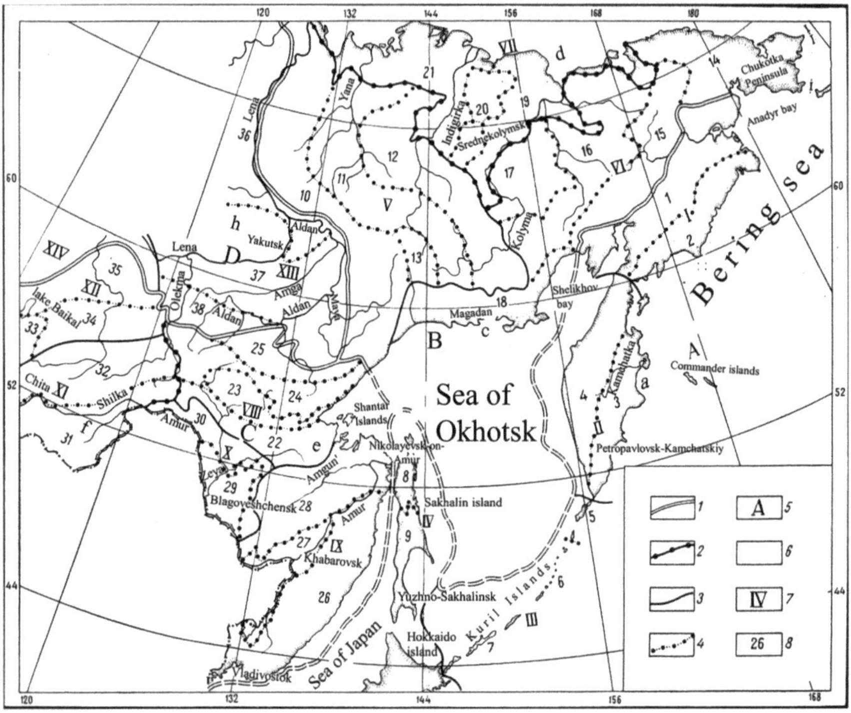

Fig. 2 Scheme of geomorphological zoning of the northwestern part of the Pacific mobile belt. Compiled by G. S. Ganeshin and Y. F. Chemekov (Krasny 1965). Geomorphological boundaries: 1—countries, 2—provinces, 3—subprovinces, 4—regions; geomorphological indices: 5—countries, 6—provinces, 7—subprovinces, 8—regions;A—Okhotsk-Anadyr geomorphological country; provinces: a—Kuril-Kamchatka, b—Sakhalin-Japan; subprovinces: I—Koryak-Anadyr, II—Kamchatka, III—Kuril, IV—Sakhalin; regions: 1—Anadyr-Penjinsky, 2—Koryakskaya, 3—East Kamchatka, 4—West Kamchatka, 5—northern group of Kuril Islands, 6—middle group of Kuril Islands, 7—southern group of Kuril Islands, 8—North Sakhalin, 9—South Sakhalin; B—Kolymo-Chukotka geomorphological country; provinces: c—Verkhoyanskaya-Chukotskaya, d—Northern Primorskaya; subprovinces: V—Verkhoyanskaya-Cherskaya, VI—Okhotsk-Chukotskaya, VII—Kolymo-Indigirskaya and Ostrovnaya; provinces: 10—Verkhoyanskaya, 11—Yano-Adychan, 12—Chersko-Moma, 13—Oymyakonskaya, 14—Chukotskaya, 15—Verkhne-Anadyrskaya, 16—Seimchano-Anyuyskaya, 17—Yukagirskaya, 18—Severo-Okhotskaya, 19—Kolyma, 20—Alazeyskaya, 21—Yano-Indigirskaya;B—Amur-Ussurian geomorphological country; provinces: e—Djugjuro-Sikhote-Alin, f—Transbaikal; subprovinces: VIII—Udsko-Okhotsk, IX—Amur-Primorskaya, X—Amur-Zeyskaya, XI—Daurskaya, XII—Verkhne-Angarskaya; regions: 22—Tukuringra-Jagdinskaya, 23—Zeisko-Udskaya, 24—Stanovy and Dzhugdzhur ridges, 25—Uchurskaya, 26—Sikhote-Alin, 27—Sredne-Amurskaya, 28—Amgunskaya, 29—Amuro-Zeyskaya, 30—Tygda-Selemdzhinskaya, 31—Shilkinskaya, 32—Vitimskaya, 33—Baikalskaya, 34—Muyskaya, 35—Mamskaya; D—East Siberian geomorphological country; provinces: g—Lenskaya; subprovinces: XIII—Lena-Aldanskaya, XIV—Verkhne-Lenskaya; regions: 36—Vilyuyskaya, 37—Amginskaya, 38—Verkhne-Aldanskaya

Table 2 Boundary system of subarctic and temperate belt coasts (Sakhalin as an example)

	Subarctic	Transition zone	Temperate
Climate zones	Northern shores of Sakhalin Island and the Northern Pacific	Eastern and southern shores of Sakhalin Island in the Sea of Okhotsk	Western shores of Sakhalin Island in the Sea of Japan
Natural conditions	Predominance of low temperatures	Frequent temperature changes from high to low after 0 degrees	Predominance of temperatures above 0 degrees
Morpholitodine-Mic factors	Rivers, waves, gravity, wind, currents, fast and pack ice, short-term period of ice formation, limited period of active ice-free sea	Rivers, gravity, vegetation, precipitation, ice, wind, long-term period of ice formation, multiple transitions of temperatures from high to low through 0 degrees	Rivers, gravity, vegetation, precipitation, wind, waves, currents, practically ice-free sea, multiple transitions of temperatures from high to low through 0 degrees
Subaerial coastal processes	Gouge, gravity, river runoff, slope processes, blowing, weathering, thermal abrasion, abrasion, fast ice rafting, accumulation	River flow, slope processes, plant and animal activity, aeolian processes, formation of fast ice and aufeis, weathering, abrasion, accumulation	Aeolian transport, blowing, slope processes, river runoff, abrasion, accumulation, active denudation in the splash and tide zone during the cold period
Sources of substances entering the coastal zone	Solid runoff, slope inputs, abrasion, beaches, Ice transport	Solid runoff, slope inputs, vegetation, shell (thanatocenoses), beaches, fast ice spreading, weathering, abrasion	Dunes, solid runoff, slope inputs, abrasion, vegetation, shell, beaches
Subaquatic landform processes on a submarine coastal slope	Waves, flows, plant and animal life, turbidity flows, longshore sediment flows, abrasion, accumulation, morphodynamics of the bottom surface beneath the landfast ice in the transition zone		

Sakhalin Island is located at the eastern edge of the Asian mainland, between 45°54′ and 54°24′ N, extending submeridionally for almost 950 km. The width of the island reaches 157 km and 50 km at the narrowest point of the Isthmus of Belt. Sakhalin is washed by the waters of the Sea of Okhotsk from the north and east. The middle and southern parts of the western coast are washed by the waters of the Tatar Strait and the northern part by the Amur estuary. The island is insignificantly distant from the mainland. The shortest distance from the mainland (Cape Lazarev) to the island (Cape Pogibi) is only 8 km. To the north and south of this point, the distance between the island and the mainland coast increases, reaching 160 km in the extreme north (Cape Elizaveta) and 250–280 km in the extreme south-west of Sakhalin (Cape Krilion).

The land areas closest to Sakhalin to the north, east and south-east of Sakhalin are even more remote. Thus, the distance between the northern tip of Sakhalin and

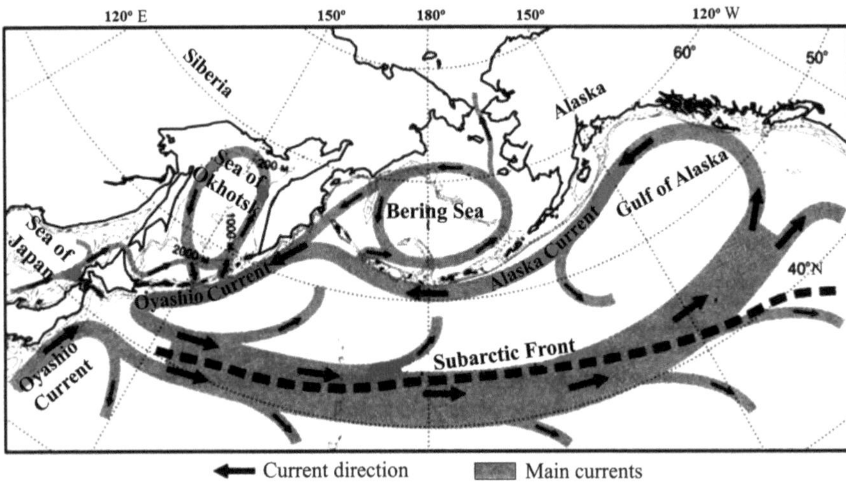

Fig. 3 General scheme of the surface water circulation of the Subarctic Pacific (Matul 2009; Marine Atlas 1953)

Table 3 Sakhalin Island coastal cartometry

	Length of coastline, km	Area, km^2
Sakhalin with lagoons and bays	4916	76,190
Sakhalin without lagoons and bays	2755	73,959
Lagoon islands	211	57
Lagoons and lakes	2161	2231

the north-western coast of the Sea of Okhotsk everywhere exceeds 350 km; the shortest distance between the south-eastern part of Sakhalin and the southern islands of the Kuril Ridge is 300–350 km, and between Sakhalin and south-western Kamchatka is about 800 km. However, Sakhalin is separated from Hokkaido Island to the south by the La Perouse Strait, which at its narrowest point is slightly over 40 km wide. Some cartometric characteristics of the coastline are presented in Table 3.

The main orographic relief elements are associated with the East Sakhalin and West Sakhalin anticlinorii and the Central Sakhalin synclinorium, which extend in a near meridional direction (Fig. 4). The meridional strike of the main structural elements generally determines the configuration of the coastline. The area of recent uplift encompasses the West Sakhalin Mountains, the East Sakhalin Mountains and the ridges of North Sakhalin (Nikitenko 1981). The area of weak uplift includes the denudation plain of northern Sakhalin and the widely developed foothills of Middle and southern Sakhalin. They have hilly or bumpy relief with intensive gully dissection and absolute elevations from 100 to 300 metres. Depressions are confined to the

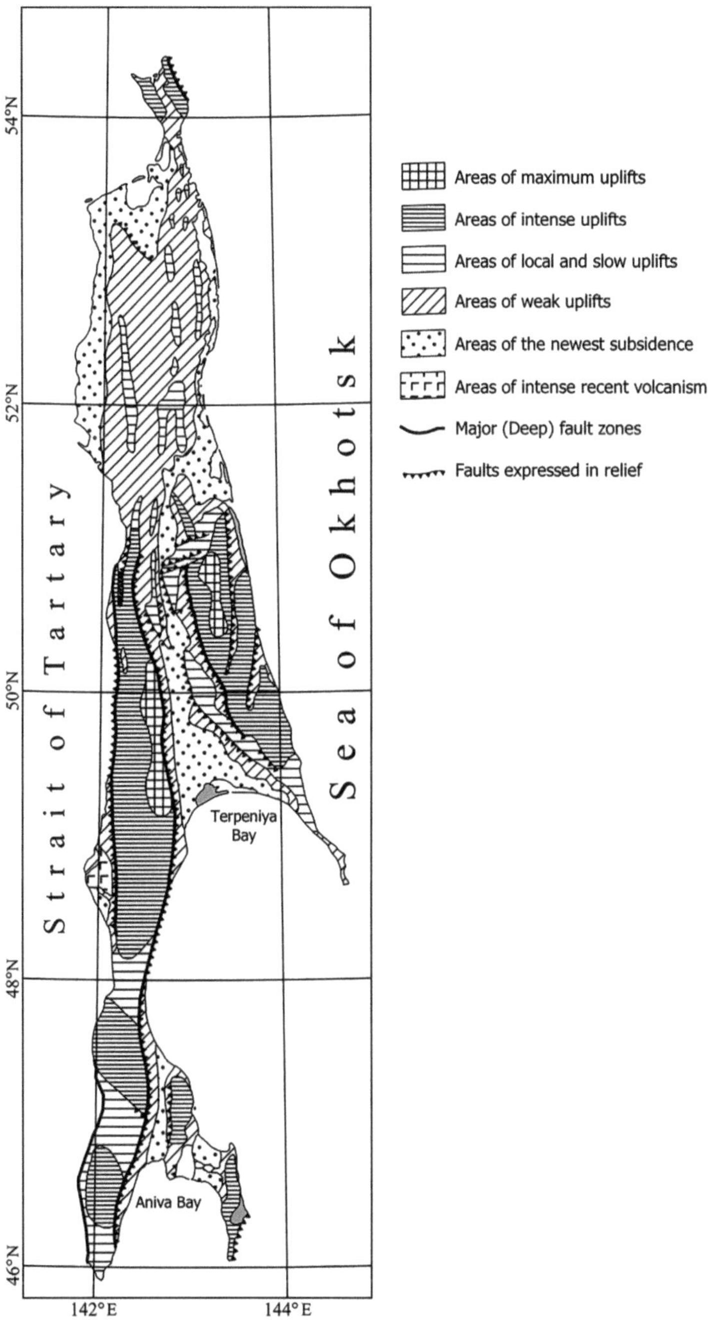

Fig. 4 Map of recent tectonics (Aleksandrov 1973)

area of neotectonic dips, in which the following geomorphological varieties can be distinguished.

Sakhalin Island's landscape is dominated by marine and alluvial plains. These include coastal plains with marine terraces (4–25 m high) made of Late Pleistocene-Holocene sediments, which border several lowlands. Alluvial, proluvial, and lacustrine-alluvial plains are also widespread, particularly on the North Sakhalin Plain and in central lowlands.

The island's climate is defined by its position in the monsoon zone, where continental and oceanic air masses meet. Stretching 1000 km from north to south, Sakhalin acts as a transitional contact zone between the subarctic and temperate zones. This results in a consistent north-south gradient in temperature and natural conditions, as well as significant climatic differences between its eastern and western coasts (Zemtsova 1966).

In general, the hydrothermal influence is manifested in the warming of the land in late autumn—winter, cooling in spring and in the first half of summer. The temperature regime of the coast is significantly influenced by the system of currents. The surface water temperature in the Tatar Strait is 3–4 °C higher than the water temperature on the eastern Sakhalin shelf (Petrov 1999). Even in the warmest month (August), the water temperature in the Sakhalin Bay is 10–12.5 °C. Similarly low coastal water temperatures were recorded at the latitude of Okha (+12 °C) with a gradual increase to the south (+13 °C at the latitude of Nogliki, +14 °C at Poronaisk, +15 °C at Yuzhno-Sakhalinsk and + 16 °C at Aniva Bay).

In the north of the Tatar Strait, the water temperature varies between +13 – +14 °C, in the south in the zone of the Tsushima Current +15 – +17 °C (Kanaev 1959). In winter time in the area of Kholmsk water temperature decreases to +1 – +2 °C. A stable ice cover is formed to the north.

The Okhotomorsk coast of Sakhalin Island has water temperatures around 0 °C throughout its entire length. In some years, when ice does not melt for a long period of time, water temperatures in the northern part of the Sakhalin shelf do not rise above +5 – +6 °C, and May temperatures are close to freezing point. The contrasting temperature regime between the western and eastern coastal areas is determined by the cooling influence of the East Sakhalin Current (especially in the warm season and the warm Tsushima Current, which increases the average temperatures of January and August and determines the date of transition of soil temperatures through zero (Fig. 5).

Due to the 1.2–2 times lower total solar radiation, ice conditions in the northern regions are comparable to those in the Arctic (Skokov 1982; Tambovsky 1999; Archikov 1989).

The Sea of Okhotsk and the northern part of the Tatar Strait are covered with ice in winter. In northern Sakhalin, ice persists for 7–8 months, somewhat less along the south-eastern coasts, and ice is extremely rare along the south-western coast of the island (Skokov 1982; Atlas of the Sakhalin Region 1967). Seven regions have been identified on the basis of the ice regime features on the island shelf (Fig. 6).

Sakhalin Bay, the Amur estuary and Nevelsky Strait are characterised by stable landfast ice, and the duration of the season with floating ice exceeds 6–7 months. Ice

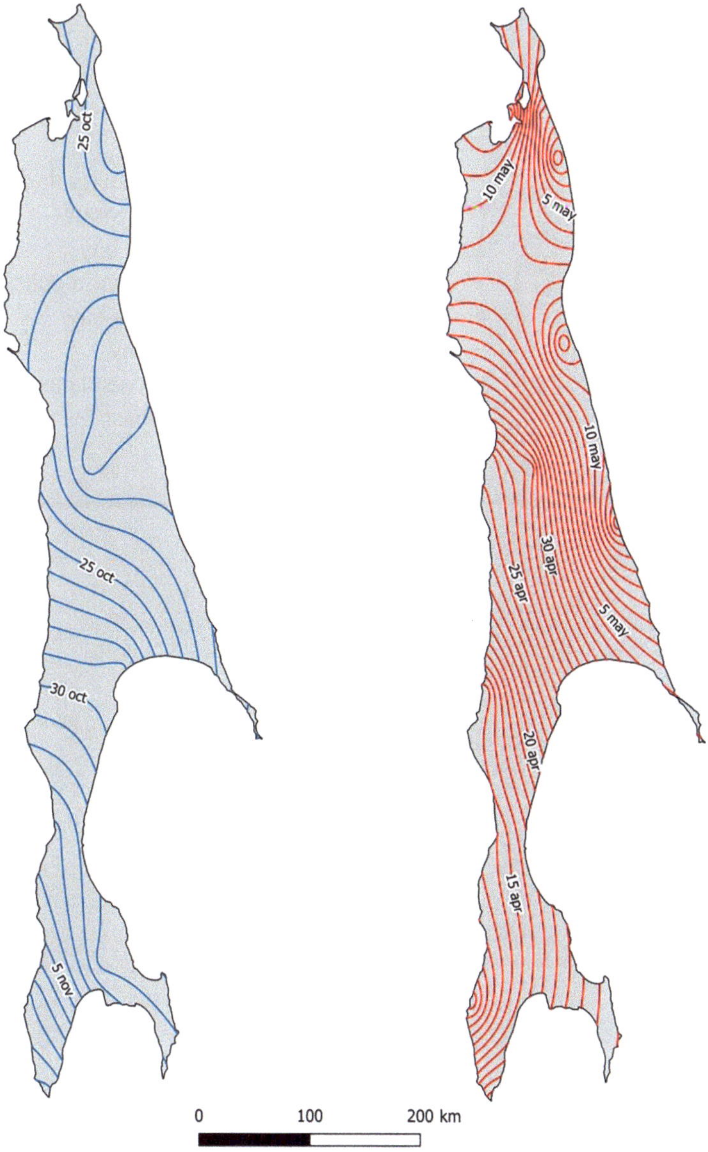

Fig. 5 Maps of soil temperature transition through zero in autumn and spring periods

conditions on the eastern Sakhalin shelf can be characterised as very severe (Tambovsky 2001; Shevchenko 2016).

Ice begins forming in the north in October-November and spreads southward. Landfast ice (stationary ice attached to the shore) typically establishes from January to February and breaks up in April-May. This landfast ice is generally unstable, especially along straight, poorly dissected coastlines, and can break up multiple

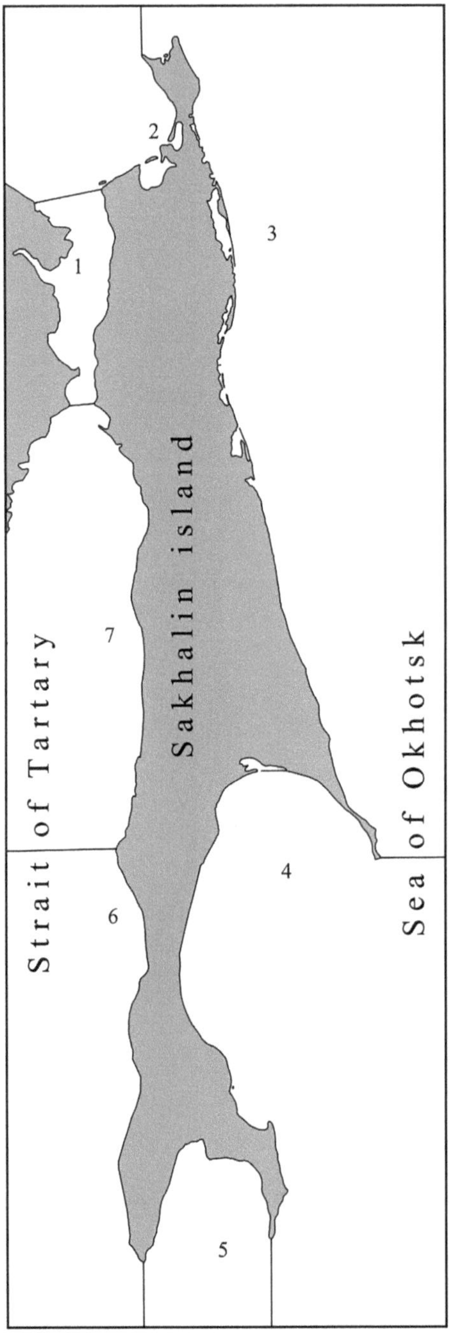

Areas

1. Amur Estuary and Nevelskoy Strait
2. Sakhalin bay
3. East Sakhalin coast
4. Terpeniya bay
5. Aniva bay
6. South-West coast
7. West coast

Fig. 6 Scheme of ice regime areas (Skokov 1982)

times during winter. It is usually 2–3 km wide, only extending to 5–6 km within sheltered bays.

Ice stability varies by region:

- **Terpeniya Bay:** Very unstable at the top, unstable along the meridional coast, and relatively unstable in the south.
- **Aniva Bay:** Has an even milder winter. The warm Tsushima Current prevents local ice formation, creating a persistent polynya (area of open water) at the top of the bay, despite ice drifting in from other areas.
- **Southwestern Coast:** Has a short ice season of 2–3 months. The Tsushima Current creates the lightest ice conditions on the shelf, preventing local ice formation entirely. Any ice present here drifts down from the northern Tatar Strait.

On the west coast of Central Sakhalin, the period with ice in the water area is 5–6 months. The first ice appears at the southern entrance to the Nevelsky Strait, which is distilled by the Amur River runoff. Nevelsky Strait in early November (Ponomareva 1982, 1989). The most severe ice conditions are observed in the southern part of Alexandrovsky Bay. In March, ice breakup begins and in April the water area of the Strait is cleared. Landfast ice on the coast is rather stable, but in the southern part it is not formed annually.

Thus, the coast of the island is characterised by very diverse ice conditions: during the winter months active ice formation takes place in the Sea of Okhotsk, and ice and landfast ice form on the shores. Ice formation is weaker in the shelf zone and on the coast of the Sea of Japan. Mean annual temperature is positive and averages 4.2 °C. Mean monthly air temperatures are positive for 8 months from April to November. The coldest month is January, with an average monthly temperature of −9.0 °C. Low temperatures in the south of Sakhalin Island are caused by the transport of cold air from the continent. The absolute minimum air temperature in the described area is observed in January-February.

The air temperature transitions through 0 °C upwards in the first decade of April. In early summer, the cooling influence of the Sea of Okhotsk takes effect. As a result, August is the hottest month in the study area. The average monthly air temperature in August was 17.5 °C (Zemtsova 1966).

In November, winter circulation conditions already prevail, and cold air intrusion occurs in the rear of cyclones. In this connection, in spite of the island position, the air temperature from October to November sharply decreases by 7–9°C.

The regime of sea level fluctuations has a complex character and depends on many hydrometeorological factors: tidal and surge phenomena, seiche fluctuations, changes in atmospheric pressure, tsunami waves, etc.

In the northern Sea of Japan, tides are irregular and diurnal, typically producing one high and one low tide per lunar day. A second, weaker high tide sometimes occurs but is not truly semi-diurnal. Tidal curves are smooth, and the amplitude between high and low tides varies significantly.

Based on data from Kholmsk, the maximum tidal range in this area is small, only 1.2–1.5 meters—the lowest around Sakhalin Island.

Seasonal sea level changes, driven by monsoons and variations in the Tsushima Current, are more significant. The cycle has two maxima (June-July and December-January) and two minima (March-April and September-October). Winter levels are generally higher than summer levels due to lower atmospheric pressure. A notable northwest-to-southeast slope of up to 0.25 m exists in winter but gradually flattens out, leveling the sea surface by the season's end.

The mean annual sea level is also subject to fluctuations. For the Sea of Japan, the main factor influencing the long-term level changes is the intensity of the Tsushima Current. Tidal fluctuations are superimposed on the fluctuations caused by baric changes during the passage of cyclones. It is not possible to trace the empirical dependence of the hydrostatic sea level rise on the atmospheric pressure drop due to insufficient instrumental observations in the area under consideration. During the passage of typhoon "Nancy" the sea level rise due to a sharp change in atmospheric pressure was 0.52 metres. (Kato 2003, 2011).

Significant storm surges on the Japanese coast of Sakhalin Island occur when deep cyclones (with surface atmospheric pressure in the centre of 950–995 hPa), mainly formed over the waters of the Yellow Sea and East China Sea, pass over the study area (Shevchenko 1994). The surges usually begin after the cyclone exits to the northern part of the Sea of Japan and reach the greatest development when the cyclone centre is located directly over the water area of the Tatar Strait or over the adjoining areas (the Amur estuary, the coast of Khabarovsk Krai, Sakhalin Island, Aniva and Terpeniya Bays). The surges end after cyclones exit to the northern or central parts of the Sea of Okhotsk. These regularities are explained by the peculiarities of the mechanism of storm surge formation in the study area: the relative deepness of the Tatar Strait determines the dominant role of the barometric factor in their formation, while the wind influence is relatively small (Shevchenko 1997).

For the Japanese-Marine stations (Uglegorsk, Kholmsk, Nevelsk), the character of this variability is practically the same, its distinctive feature being the presence of a pronounced maximum of surge intensity in April and in the autumn-winter period (October-December). The average height of the surge level for these months at all stations is in the interval from 30 to 40 cm. The lowest intensity of storm surges is noted in June-July, when the average heights do not exceed 20 cm.

Seasonal variations in hydrological conditions and phenomena in northeastern Sakhalin are significant and are characterised by high values in the autumn-winter period (October-January) and low values in the spring-summer period (March-August), which is fairly typical of coastal stations in the Sea of Okhotsk (Poezhdalova 1997). In February and September, a transient regime is observed, which is characterised by rapid changes in the mean level. The maximum differences in the annual variations between the absolute maximum (January) and minimum (April) are about 15 cm. At the same time, significant deviations from the mean multiyear values are observed in different years, as evidenced by the curves of extreme values of mean monthly level heights; the difference between the absolute maximum and minimum reaches half a metre. The formation of the winter maximum in the Sea of Okhotsk is explained by the influence of the winter minimum of atmospheric pressure in the North Pacific Ocean—the Aleutian Depression (Sedaeva 2001).

Long-term data from Nabil Bay shows that typical sea level fluctuations are small, averaging 8–10 cm in autumn/winter and 4–5 cm in summer. However, these minor variations are occasionally interrupted by anomalous storm surges. These extreme rises in water level are caused by cyclones or typhoons, which drive surges with strong winds and a sharp drop in atmospheric pressure.

The maximum value of the surge level observed during the above-mentioned observation period was 85.9 cm (November 1995), the second largest surge with a height of 85.0 cm was recorded in October 1986. Two more surges exceeding the mark of 75 cm and classified as dangerous events according to the existing classification were observed in September 1961 (75.5 cm) and May 1996 (75.1 cm). The average frequency of storm surges with a height of at least 50 cm is about 4 cases per year.

3 The Significance of Coastal Wave Exposure on Inland Seas and Open Ocean Coasts

The wind regime is determined by the monsoon type of air circulation, which is characterised by a seasonal change of the prevailing winds opposite in direction. The general properties of the wind field over Sakhalin are well studied (Atlas of Pacific Ocean waves and winds 1966; Atlas of waves and winds of the Sea of Japan 1963; Koshinsky 1976). The peculiarities of the wind field in winter in this area are due to the location of the island at the boundary between the zones of influence of the oceanic (Aleutian depression) and continental (Siberian anticyclone) seasonal atmospheric centres. During the summer period, the North Pacific anticyclone and the Asian Depression create south and south-east air flows over Sakhalin (Zemtsova 1966). Stable, strong winds of north-west directions, with the movement of dry and cold air masses formed in the northern part of Siberia, are formed in the cold period (Zemtsova 1966). When working on the coast, we widely used the Popov-Sovershaev wind-energy method of calculating the nanomotive force (Popov 1981). Therefore, the wind information was used in the work quite extensively. However, a generalisation of the wind energy potential of the Sakhalin Oblast by P. D. Kovalev et al. allows us to present this information in a condensed form (Kovalev 2015).

As noted by the authors of this summary, all stations are characterised by a pronounced annual course with almost cyclic variations (Fig. 7), with fairly strong (4–8 m/s) north and north-west winds prevailing over Sakhalin in winter.

During summer, winds from the south and southeast prevail, with typical speeds of 1–4 m/s. The highest average wind speeds are recorded on the island's southwest coast (e.g., at Cape Krilion and Nevelsk). Overall, summer monsoon winds on Sakhalin are significantly stronger than those on the Kuril Islands, except for the northern Kurils, where the monsoon's influence is absent (Kovalev 2015).

Fig. 7 Monthly mean wind speed vectors characterising the seasonal variability of the wind field for eastern (Odoptu-Vzmor'e) and western (Pogibi-Korsakov) regions of Sakhalin (Kovalev 2015)

Table 4 Number of days with wave heights not exceeding the given values, according to observations at the Floating drilling rig (FDR) (Afanas'ev 2015a, b)

Threshold heights, m	July-August	September	October-November
Northern region			
≤0.5	6.9	3.1	1.8
≤1.0	19.2	11.9	9.7
≤2.0	26.1	20.6	15.7
≤3.0	30.3	27.6	23.7
South region			
≤0.5	5.7	3.1	1.9
≤1.0	15.7	11.5	8.7
≤2.0	23.8	21.2	14.7
≤3.0	30.0	27.0	23.2

3.1 *Wind-Wave Conditions in Western Sakhalin Island*

The coastal waters of northwestern Sakhalin are a unique natural laboratory. The absence of complex wave patterns makes it an ideal location for using wind data to analyze water movements.

In the Tatar Strait, storms are caused by cyclones and occur mainly in the cold season. Wind speeds during these storms can exceed 40 m/s. The area averages 45 storm days per year, with winds surpassing 15 m/s. The typical monsoon pattern is frequently interrupted by these cyclones in autumn and winter, which bring violent storms and heavy precipitation.

Wave activity in the southwestern Sea of Okhotsk is directly influenced by the region's atmospheric circulation patterns. The storm and calm conditions for the northern and southern parts of this area (divided at the 52nd parallel) are detailed in Tables 4 and 5.

For the summer period, weak and moderate winds of southeastern and southern directions prevail over the water area. Accordingly, the wave regime is characterised by the prevalence of waves with wave heights up to 1 m. The number of days with such waves reaches 16-19 per month. Wave heights of more than 2 m are not more frequent than 5–7 days, while waves of more than 3 m are observed on average 1 day per month. In September, the frequency of weak (1 m and less) swell is about 12 days, swell of more than 2 m is observed for 9–10 days during the month, and more than 3 m—for 2–3 days.

The October-November period is characterised by the formation of winter monsoon conditions over the Sea of Okhotsk. Wave conditions at this time are predominantly determined by north-westerly quarter winds, whose speeds increase markedly compared to September. A swell of 1 m or less in October-November was recorded for 2 days. A swell of more than 2 m is observed for about 16 days during the month and a swell of more than 3 m for about 9 days.

Tables 4 and 5 show that there are no significant differences in wave characteristics between the northern and southern areas. The wave heights most likely to occur

Table 5 Mean and longest duration (hr.) of cases with excitement not exceeding specified values, as observed at the Floating drilling rig (FDR) (Miller 1960; Reimnitz 1965)

Threshold heights, m	July-August	September	October-November
Northern region			
Calm	8/34	6/24	5/22
≤. 0	36/152	27/112	22/114
≥2.0	10/36	18/102	24/110
South region			
Calm	8/34	6/24	5/22
≤1.0	33/144	25/103	21/78
≥2.0	13/40	16/88	27/101

in winds of different directions and speeds are shown in Fig. 6. They increase naturally with increasing wind strength. Wave heights increase most rapidly with increasing winds of open rhumb winds—C, NE and W directions. Obviously, this is connected with the arrival of quicksand from the Sea of Okhotsk (Fig. 8).

Winds from the southeast are an exception. The wave heights they generate increase at a much slower rate as the wind speed rises.

The frequency of storm surges grows significantly from summer to winter. They make up 30% of events in the summer, increasing to 50% by September and 80% by November. Similarly, the proportion of severe storms with waves exceeding 4 meters is small in summer (0–7%) but rises substantially to 30–40% in autumn and winter.

The highest wave height ever recorded at the Piltun station was 7.8 meters.

Catastrophic tsunami waves are the most threatening type of marine natural disasters in the Russian Far East. However, on the coast of Sakhalin, particularly in the north-east, they are much less dangerous than on the Pacific side of the Kuril Islands and Kamchatka. A number of records were obtained at the gauge station in Nabil Bay, including a record of one of the strongest tsunamis on the coast of the Sea of Okhotsk, the Chilean tsunami (May 1960). The wave height was 80 cm. Numerical modelling (Shevchenko 1990) showed that the wave attenuation coefficient from the open coast to the location of the tide gauge is 1.3, which means that the wave height of the Chilean tsunami was probably about 1 m in the area of the open coast.

3.2 *Wind-Wave Conditions in the Southern Part of the Sea of Okhotsk*

According to data of HMS Kurilsk, the storm impact on the Okhotomorsk shores is weakened in spring and summer. In the autumn-winter season, the repeatability of high waves increases significantly (Fig. 9).

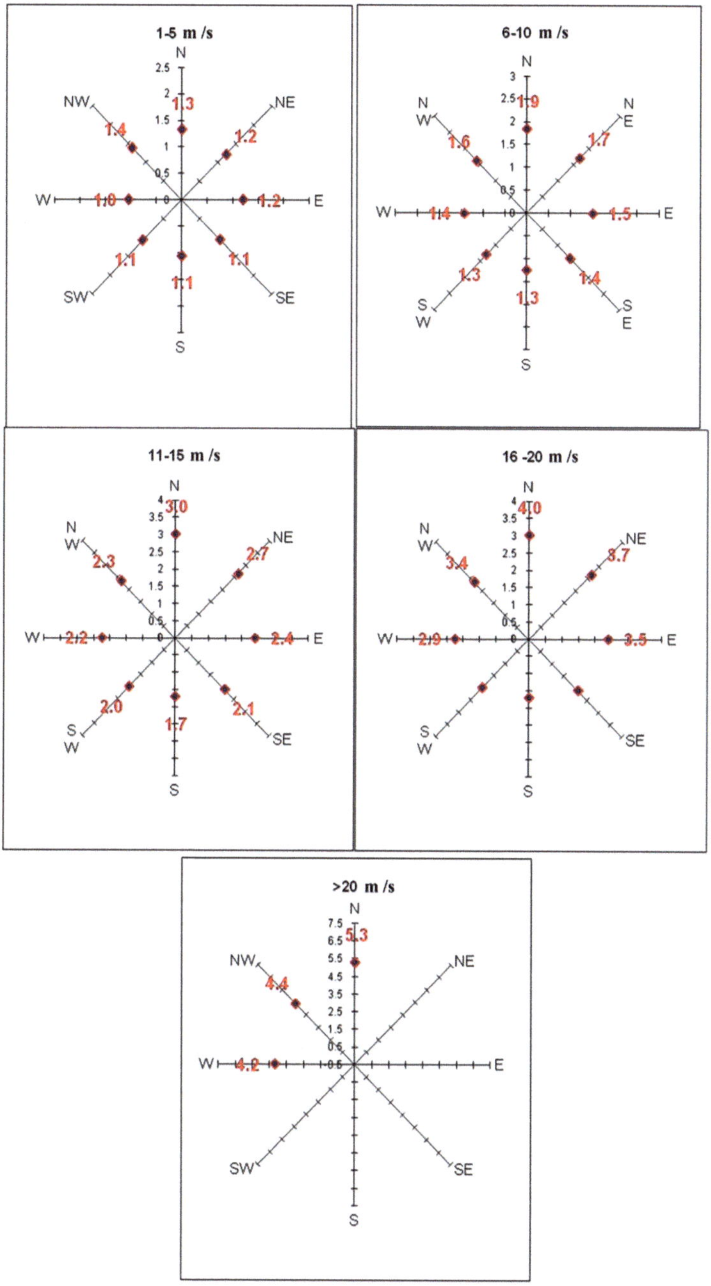

Fig. 8 Most probable wave heights at different wind speeds and directions (Afanas'ev 2015c)

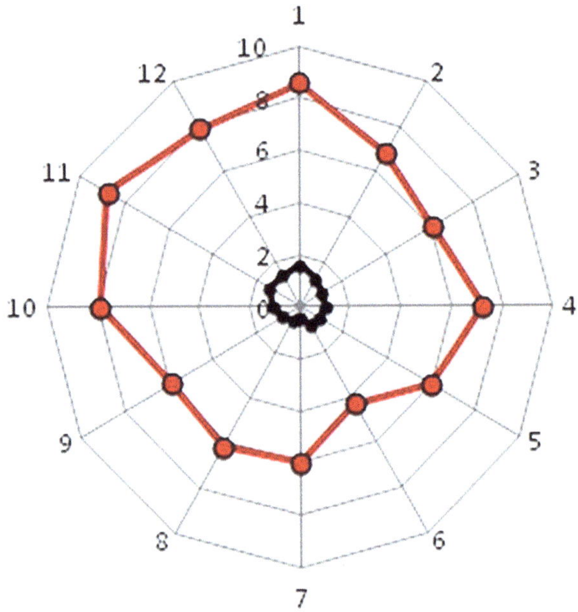

Fig. 9 Monthly average wave heights (average—black line, and maximum—red line). [http://portal.esimo.ru/dataview/viewresource?resourceId=RU_RIHMI-WDC_2370&parameter=waves&sea=okhot&filter=workingstations]

Waves of 5 m and more make up 6–7% of the total number of waves. The strongest waves are produced by cyclones (typhoons). Typhoons usually enter the southern part of the Sea of Okhotsk in September, travelling along the Kuril Ridge. NW winds and waves prevail in summer, NW, NE and SE winds and waves are more frequent in autumn, NE in winter, and SW and NE in spring. In general, wave directions from NW to NE prevail, which determines a rather high intensity of coastal processes in the bay (Dunayev 2019).

Two zones of intense wave intensity associated with the convergence zones of cyclone trajectories located in the west and east of the North Pacific Ocean were identified in the North Pacific Ocean (Kaplunenko 2011; Polyakova 2004, 2007). The wave intensity corresponds to the seasonal course of the atmospheric processes: the largest in winter, somewhat weaker in autumn, much weaker in spring and weakly pronounced in summer (Fig. 10).

The study employs Iturup Island to investigate contemporary coastal morpholithogenesis and aeolian accumulation—a set of relief-forming processes referred to as nearshore transport that operate over palaeogeographic timeframes.

Variations in modern coastal morpholithodynamics are most evident in the morphometric parameters of beach megafestoons. A detailed analysis of these specific morpholithodynamic structures is provided in a subsequent section.

Here we only note that the morphometric parameters of oceanic megafestons are more than twice as large as those of the Okhotomans (Figs. 11 and 12).

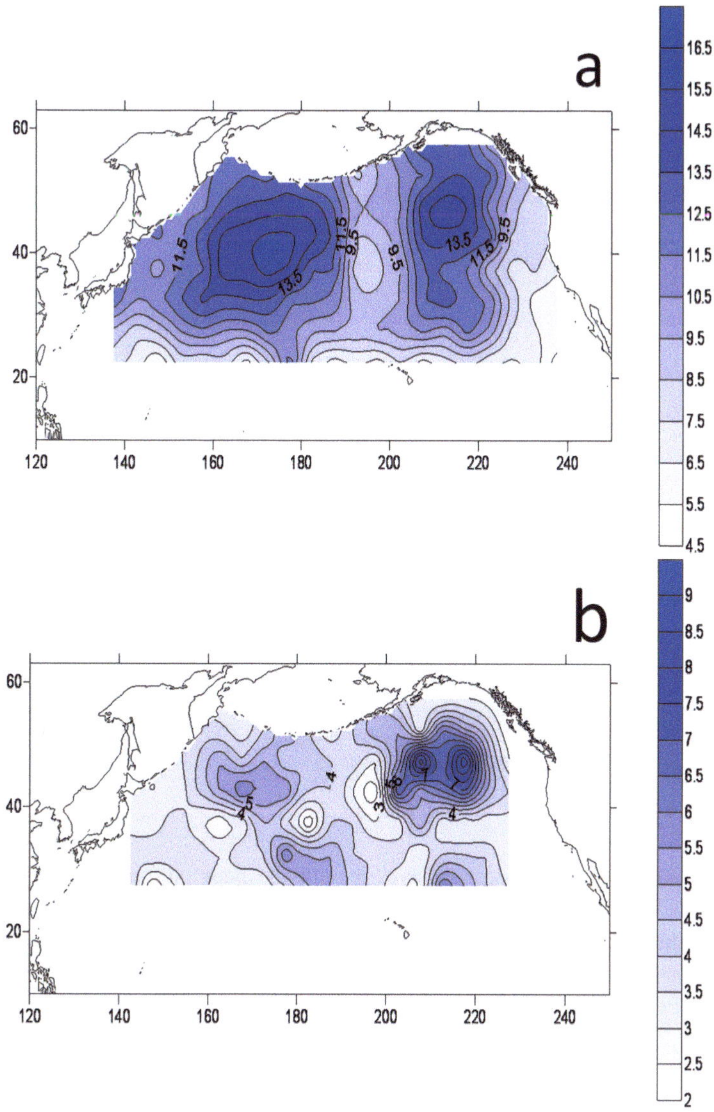

Fig. 10 Repeatability of wave heights of 6 m and more, %: (**a**) from September to March, (**b**) from April to August (Dunayev 2019)

This section does not provide an analytical review of coastal aeolodynamics, except to note that the problem of the cyclicity of aeolian morpholithogenesis on the coast has been actively discussed to date (Vykhovanets 2003).

In our opinion, the cyclicity of aeolian morpholithogenesis on the coast is related to changes in the sediment balance in the coastal zone, determined mainly by sea-level fluctuations. The main regularities of aeolodynamics on the shores of the Far

Fig. 11 Okhotomorsk megafestons (L = 400–600 m)

Fig. 12 Pacific megafestoons (L = 1200–1500 m)

Eastern seas in the Holocene have previously been summarised as follows (Afanas'ev 1992).

In coastal areas characterised by large volumes of debris removal from the land, the development of aeolian accumulation processes is associated with a decrease in sea level. For example, in the Viakhtu Bay area of northern Sakhalin, the formation of a 2.0–2.5 m thick aeolian strata on the surface of a 20 m alluvial-marine Upper Pleistocene terrace began simultaneously with the formation of a young eutrophic peat bog on the estuarine-lagoonal drainage of the bay (440 ± 83 cal. l.n. FESU -187; 440 ± 85 cal. l.n. FESU -178).

On the coast, where sediment supply to the coastal zone is provided predominantly by marine hydrodynamic processes, aeolian deposits are formed during the period of increased scour during sea-level rise. The thick dune deposits of the Sakhalin coast of the Amur Liman began to form on the accumulative segments of abrasion-accumulative systems precisely at the beginning of the second phase of wave accumulation (4615 ± 579 cal. l.n. FESU -109).

In Russia, the coasts studied mainly along the estuaries of large rivers laid the foundation for the regressive theory of dune formation (Korotky 2007; Korotkiy 1995a). And on the coasts with active coastal erosion, the opinion about the activation of aeolian relief formation during sea level rise was formed (Badyukova 1997).

To be fair, it should be noted that recently there has been an understanding of our view that aeolian processes in the coastal zone are controlled by the parameters of the sediment budget during sea-level fluctuations of different sign (Afanas'ev 1992; Badyukova 2015; Aagaard 2007).

Some young forms of aeolian-coastal accumulation probably do owe their origin to changes in hydrodynamic activity over the last few decades, provided that these processes are inherited at the development sites. This applies in particular to dunes on the coasts of Sakhalin Bay, the Tatar Strait and north-east Sakhalin. Modern aeolian deposits are observed here at elevations up to +25 metres. The high mobility of sandy material explains the presence of Early-Middle Holocene shoreline aeolian deposits on the coast.

We also consider the issues of aeolian morpholithogenesis associated with wave processing of pumice-pyroclastic material during explosive eruptions in the Holocene and Neopleistocene. Aeolian covers in volcanic centres have been studied for quite a long time (Porter 1997; Botvinkina 1974). However, in our case, we focus on the formation of dunes as a result of excessive sediment of pyroclastic origin in the coastal zone.

On Iturup Island (Kuril Ridge), four large collapse calderas were formed in the late Neopleistocene-Holocene, during the formation of which about 450 km^3 of predominantly dacitic pyroclastics were erupted (Melekestsev 2005). This resulted in the influx of hundreds of millions of cubic metres of pumice pyroclastic material and tephra into the Pleistocene-Holocene coastal wave recycling zone.

We have established that in the middle-late Holocene, after the sea level rise to levels close to the present day, at least two large pyroclastic series were formed in the central part of the island, the last of which is dated 1050 ± 70 cal. yr. BP—LU-8685. Volcano-tectonic dislocations with a layer of beach material were found at the base of the section of the high seashore scarp. The shell material from the ancient beach sediments ^{14}C 6220-6420 cal. yr. BP (Lu-9223) and 7550–7690 cal. yr. BP (Lu 9241) (Afanas'ev 2019) allows us to attribute these dislocations to the Middle Holocene, the period of maximum sea level rise (Fig. 13).

According to the predecessors, the dunes with a series of shallow buried soils on the Okhotomorskaya side of the Vetrovaya Isthmus are compared with the dunes of the second generation of Kunashir Island ^{14}C 1214 ± 79 cal. yr. BP (GIN7884) (Korotkiy 1995b). On the Pacific side of the isthmus, a ridge of Little Ice Age dunes extends (Razhigayeva 2005).

Our studies have shown that the Okhotomorsk coast of the Vetrovaya Isthmus is a basement terrace developed in pumice and overlain by aeolian sediments up to 25–30 m thick. Surface gradients of the pumice benches buried beneath the dunes in the 300–400 metre strip of coastline are 1.4–1.6° (Fig. 14a, b, e).

In the middle part of the aeolian deposits there is an ash interbed represented by white siltstone. The formation of this layer probably occurred between 250 BP (14C-dates 250 ± 45 AA-20942), and 800 BP (^{14}C-dates 800 ± 80 BP, AA-20941;

Fig. 13 Sub-vertically stratificated layers of pumice-pyroclastic deposits with interlayer of sea beach sediments in the coastal ledge on the south-western boundary of the Okhotsk sea coast of Vetrovoy Isthmus

880 ± 40 BP, GIN-7344). The age of the underlying sediments is consistent with the age determined earlier (Fig. 14e).

The situation is somewhat different on the oceanic coast (Fig. 14c, d).

At the base of the dunes lies a low thickness peat bog, the formation of which ended about 1000 years ago (LU-9228). This period corresponds to the formation of humus ash interlayers on the ocean and seashore (LU-9219, LU-8685), overlain by overweathered ashes up to 10 m thick on the seashore (Fig. 15). Thus, the formation of younger oceanic coastal dunes is obviously related to the reworking of pyroclastic material that entered the coastal zone about 1000 years ago. kyr.

Overlying the soils (LU -9220, 680-880 calendar years B.P.) are aeolian sands up to 8 m thick. The second cycle of aeolian morpholithogenesis here is probably associated with the erosion of the first generation dunes.

4 Conclusion

The coastal zone is a vital resource for humanity, serving as a cradle of civilization whose ecological and functional integrity is essential for the sustainable development of coastal regions. Although the fundamental principles of coastal morpholithodynamics are universal, significant regional variations—such as those in morpholithogenesis—also exist.

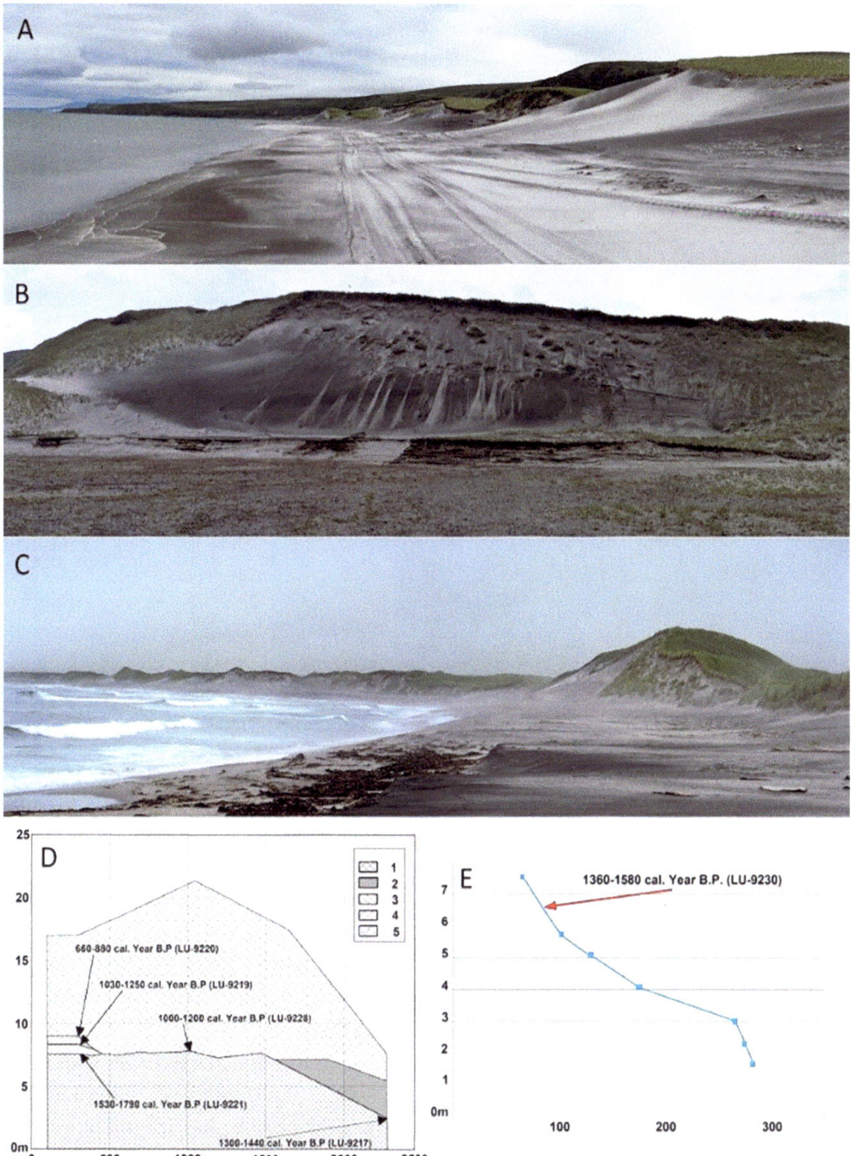

Fig. 14 Dunes of the contact zone. (**a**, **b**) Okhotomorskiy dunes (H max = 23.1 m (modern); 34.0 m), (**c**) Pacific dunes H max = 17.8 m; (**d**) longitudinal section of the oceanic coastal scarp with the location and radiocarbon dating of sediments, (**e**) position of the dune base with radiocarbon dating on the pumice basement of the marine terrace

Fig. 15 Volcanic ash correlated with dunes on the coast of Iturup Island

The Russian Far Eastern seas, which form part of the North Pacific's broader cyclonic circulation, span multiple physiographic zones: Arctic, subarctic, and temperate. Predicting coastal evolution and understanding variations across these zones and sectors is an urgent and highly practical scientific challenge. Industrial development of these coasts began during a period of declining coastal dynamics and clear accretionary trends. However, by the mid-twentieth century, intensified erosion was observed across all these seas. This occurred amid inadequate safety protocols and a historically complacent attitude toward marine hazards.

The expansion of oil and gas infrastructure markedly altered this situation, generating—and continuing to produce—vast amounts of engineering-geomorphological data. However, the thematic and geographic fragmentation of applied research has hindered synthesis of this information. Systematizing these findings could provide new momentum for geomorphology as a foundational discipline within geography. A further complication is that many dynamic features observed along these shores defy explanation by conventional theory, since subarctic coastal systems have historically been overlooked as distinct objects of study.

References

Vladimirov, A. T. Studies on the dynamics and morphology of the coasts of the Sea of Okhotsk and the Sea of Japan / A. T. Vladimirov, V. S. Medvedev // Proc. of the Oceanographic Commission of the USSR Academy of Sciences. Oceanographic Commission of the USSR Academy of Sciences. - 1959a. - T. 4.

Vladimirov, A. T. Morphology and evolution of the lagoonal shore of Sakhalin Island / A. T. Vladimirov // Proc. of the Institute of Oceanology of the USSR Academy of Sciences. Institute of Oceanology, USSR Academy of Sciences, vol. 46. - M.: ANS SSSR, 1961. - P. 145-171.

Vladimirov, A. T. Evolution of the West Kamchatka shore in the Quaternary time / A. T. Vladimirov // Proceedings of the Oceanogr. Commiss. ACADEMY OF SCIENCES OF THE USSR. - 1959b. - VOL. 4. - P. 202-214.

Medvedev, B. C. Some issues of dynamics and morphology of the western coast of Sakhalin in connection with the construction of small fishing ports / V. S. Medvedev // Proc. of the Oceanogr. Commission. Oceanographic Commission. - 1959. - VOL. 4. - P. 3-12.

Medvedev, V. S. Some features of the structure and dynamics of the western coast of Sakhalin Island / V. S. Medvedev //. - Proc. Oceanogr. Commission of the USSR Academy of Sciences. - T. 8, 1961. - P. 65-84.

Ionin, A. S. Shores of the Bering Sea / A. S. Ionin. - M.: ANS SSSR, 1959. - 358 p.

Ionin, A. S. Relief of the World Ocean shelf / A. S. Ionin. - M. Nauka, 1992. - 255 p.

Kaplin, P. A. Some features of the Kuril-Kamchatka zone coast relief in connection with the tsunami problem / P. A. Kaplin, A. S. Ionin // Bul. of the Seismology Council. Council on Seismology. - 1961. - №. 9. - P. 74-86.

Kaplin, P. A. About some peculiarities of the lagoons of the north-eastern coast of the USSR / P. A. Kaplin // Proc. of the Oceanographic Commission of the USSR Academy of Sciences. Oceanographic Commission of the USSR Academy of Sciences. - 1957. - VOL. 2. - P. 104-110.

Kaplin, P. A. Fiord coasts of the Soviet Union / P. A. Kaplin // M. Izd-wo. ACADEMY OF SCIENCES OF THE USSR, 1962. - 188 p.

Zenkovich, V.P. Pacific Ocean shores / V.P. Zenkovich, A.S. Ionin, P.A. Kaplin, B.C. Medvedev. - Moscow: Nauka, 1967. - 376 p.

Zenkovich, V. P. Dynamics and Morphology of Sea Shores. Ch I: Wave Processes / V. P. Zenkovich. - M.-L.: Marine Transport, 1945. - 496 p.

Zenkovich, V. P. K geomorphology of the western coast of Kamchatka / V. P. Zenkovich, O. K. Leontiev, L. G. Nikiforov, S. A. Lukyanova // Geomorphology and lithology of the coastal zone of seas and other large water bodies. - M.: Izd-vo Nauka, 1971a. - P. 3-6.

Zenkovich, V. P. Marine geomorphology: Terminological reference book: Coastal zone-processes, concepts, definitions [Text] / V. P. Zenkovich. - Mysl, 1980. - 280 p.

Zenkovich, V. P. Some observations on the morphology of the estuaries of small rivers of the lowland Kamchatka coast / V. P. Zenkovich // Izvestia VGO. - 1953. - Vol. 85. 5. - P. 598-602.

Zenkovich, V. P. Features of the relief formation and modern sediments of the coastal zone of the Far Eastern seas of the USSR / V. P. Zenkovich, K. Leontiev, A. T. Vladimirov et al. - Moscow: Nauka, 1971b. - 181 p.

Budanov, V. I. On the signs of modern vertical movements of the shores of the Far Eastern seas / V. I. Budanov, A. S. Ionin, V. S. Medvedev // Proceedings of the Council for the Study of the Quaternary Period. - Moscow: Izd. of the USSR Academy of Sciences, 1961. - T. 1. - P. 132-142.

Gershanovich, D. E. Geomorphological and lithological studies of the accumulative shore of the Sea of Okhotsk near Okhotsk / D. E. Gershanovich, E. K. Zabelina // Proc. of GOIN. GOIN. - 1957. - Issue. 34. - P. 93-141.

Zabelina, E. K. Migration of the Okhota River breakwater and its influence on the shore dynamics / E. K. Zabelina // New studies of the shores of seas and reservoirs. Proc. Oceanographic Commission. - 1961. T. 12. - P. 67-72.

Leontiev O. K. Geomorphology of sea shores [Text] / O. K. Leontiev, L. G. Nikiforov, G. A. Safyanov - M.: Izd-vo MSU, 1975. - 336 p.
Leontiev, O. K. Morphology and genesis of Schastya Bay (Sea of Okhotsk) / o-va K. Leontiev, V. A. Lorgus, S. A. Lukyanova // Bulletin of Moscow University: Geography. - 1967. - T. 22.
Leontiev, O. K. Novel vertical movements and dynamics of the shores of the mainland coast of the Sakhalin Gulf of the Sea of Okhotsk / Ova K. Leontiev, V. A. Lorgus, S. A. Lukyanova // Development of sea shores in the conditions of oscillatory movements of the Earth's crust. - Tallinn. - 1965. - P. 129-137.
Kanaev, V. F. Geomorphological observations on the Kuril Islands / V. F. Kanaev // Proc. of IOAS. IOAN. - VOL. XXXII, 1959. – P. 215-231.
Pavlidis, Yu. A. Some features of the formation of modern coastal sediments within the volcanic archipelago: on the example of the Kuril Islands / Yu. A. Pavlidis. - Moscow: Nauka, 1966. – 112 p.
Nevesskiy, E. N. Sedimentation Processes in the Coastal Zone of the Sea / E. N. Nevesskiy. - Moscow: Nauka, 1967. - 255 p.
Miller, Don J. The Alaska earthquake of July 10, 1958: giant wave in Lituya Bay / Miller, J. Don // Bulletin of the Seismological Society of America. – 50. 2. – 1960. – P. 253–265.
Reimnitz, E. Effects of the Alaska earthquake and tsunami on recent deltaic sediments / E. Reimnitz, N. F. Marshall // Journal of Geophysical Research. – 70. 10 – 1965. – P. 2363–2375.
Pararas-Carayannis, G. A study of the source mechanism of the Alaska earthquake and tsunami of March 27, 1964: Part I, Water waves / G. Pararas-Carayannis // Pacific Science. – 21. – 1967. – P. 301–310.
Wilson, B. W. The tsunami of the Alaskan Earthquake, 1964: Engineering evaluation Vol. 25/ B. W. Wilson, A. Torum. – US Army Coastal Engineering Research Center, 1966. – 449 p.
Archikov, E. I. Abrasion factor of sedimentary material inflow into the Sea of Okhotsk / E. I. Archikov, P. F. Brovko, V. F. Rybakov, Y. D. Shuisky // Modern sedimentation and Quaternary morpholithogenesis of the Far East. - Vladivostok: Far Eastern Branch of the USSR Academy of Sciences. 1982. - P. 165-177.
Archikov, E. I. Features of the Pacific Ocean shore relief formation in the subarctic and temperate climatic belts / E. I. Archikov // Studies of global factors of climorphogenesis of the Far East: a collection of scientific works - Vladivostok: Far Eastern Scientific Centre of the Academy of Sciences of the USSR, 1979. - P. 70-76.
Archikov, E. I. Problems of theoretical and applied geomorphology of the Far East seashores / E. I. Archikov. - Vladivostok: Publishing house of the Far Eastern State University, 1985. - 124 p.
Kulakov, A. P. On the development of the coastlines of the Sea of Okhotsk and the Sea of Japan in the Quaternary / A. P. Kulakov // In the book: Issues of Quaternary Geology. - Khabarovsk, 1966. - P. 177-176.
Kulakov, A. P. On the development of the coastlines of the Sea of Okhotsk and the Sea of Japan in the Quaternary / A. P. Kulakov // In the book: South of the Far East. - M., 1972. - P. 307-339
Kulakov, A. P. Quaternary coastlines of the Sea of Okhotsk and the Sea of Japan / A. P. Kulakov. - Novosibirsk. Nauka, 1973. - 188 p.
Hershberg, L. B. Ancient shorelines of the postglacial transgression on the shelf of the Sea of Japan and the Sea of Okhotsk / L. B. Hershberg, A. A. Ryazantsev, L. G. Guskov et al. //. - In the book: Sea and Ocean Level Fluctuations for 15000 Years. M., 1982, P. 196-207.
Korotkiy, A.M. Sea level fluctuations of the Sea of Japan in the Quaternary period (north-western shelf) / A.M. Korotkiy. - In the book: Sea Level Changes. - M., 1982. - P. 104 -114.
Korotkiy, A. M. Exogenous geomorphological systems of sea coasts [Text] / A. M. Korotkiy, G. I. Khudyakov. - Moscow: Nauka. 1990. - 216 p.
Kravtsov, V. D. Some peculiarities of the dynamics of tidal river mouths of the north-western coast of the Sea of Okhotsk / V. D. Kravtsov // Proc. of GOIN. GOIN. - 1980. - Issue. 159. - P. 71-84.
Kravtsov, V. D. Some peculiarities of development of tidal estuaries of the north-west coast of the Sea of Okhotsk / V. D. Kravtsov // Proc. of GOIN. GOIN. - 1979. - Issue. 143. - P. 102-131.

Veinbergs, I. G. Ancient coastlines on the shelf and in the coastal land strip of the south-western part of the Sea of Okhotsk / I. G. Veinbergs // Shelves, Nature and Resources. - Л.: 1975. - P. 55-55.
Veinbergs, I. G. Ancient sea shores of the USSR / I. G. Veinbergs. - Diss. Doktor. geogr. sciences. - Riga: University of Latvia, 1990. - 428 p. (core.ac.uk/download/pdf/71748975.pdf)
Grabkov V. K. Morphogenetic types of relief of the Kuril Islands / V. K. Grabkov, A. A. Ischenko // Relief and volcanism of the Kuril island-arc system. - Vladivostok: Far Eastern Scientific Centre of the Academy of Sciences of the USSR, 1982. - P. 13-24.
Grabkov, V. K. Marine terrace series as indicators of transgressions and indicators of the character of movements of the transition zone from the Asian continent to the Pacific Ocean / V. K. Grabkov // Palaeogeographic aspects of changes in natural conditions of Siberia and the Far East. - Novosibirsk: Nauka. Siberian Branch, 1969. - P. 151-15.
Gurieva, Z. I. Geological and geomorphological study of sea shallows and shores on the basis of aerial photographs / Z. I. Gurieva, K. M. Petrov, N. S. Ramm, V. V. Sharkov. - L.: Nauka. Leningrad Branch, 1966. - 372 p.
Brovko, P. F. Formation of lagoonal shores of the Far Eastern seas under temperate climate conditions / P. F. Brovko // Benthos and conditions of its existence on the shelf zones of Sakhalin. - Vladivostok: Far Eastern Scientific Centre of the Academy of Sciences of the USSR, 1985. - P. 4-19.
Bobykina, V. P. Modern abrasion process on the ocean shores of Kamchatka / V. P. Bobykina // Mankind and coastal zone of the World Ocean in the XXI century. - MOSCOW: GEOS. 2000 - P. 373-377.
Bobykina, V. P. The main factors and processes of formation of the oceanic coast of East Kamchatka / V. P. Bobykina // Coastal zone of the sea. - Moscow: Nauka. 1981. - P. 117-125.
Barkov, L. K. Dynamics and morphology of the coastal zone of the North-Eastern part of Sakhalin / L. K. Barkov // Izv. VGO. - 1985. - T. 117-5. - P. 525-531.
Denisov, S. V. About the coastal-marine gold placers of the southern coast of the Sea of Okhotsk / S. V. Denisov // Geology and Geophysics. - 1967. - № 4. - P. 28-34.
Denisov, S. V. Types of the southern shores of the Sea of Okhotsk, coastal-marine sediments and their mineralogy / S. V. Denisov // In the book: Problems of the Quaternary Period Study. - Khabarovsk, 1966. - P. 174-175.
Polunin, G. V. Data on the rate of abrasion processes on the shores of Sakhalin / G. V. Polunin, V. A. Buzlaev, V. A. Konovalenko // Problems of geography of the coasts and shelf of the Far Eastern seas. - Vladivostok: DVGU, 1977. - P. 134-144.
Polunin, G. V. Relief dynamics and methods of research of the isolated marine coastal system (on the example of the Tyulenii Island polygon) / G. V. Polunin, V. A. Buzlaev, V. A. Konovalenko. - Yuzhno-Sakhalinsk: IMGiG FEB AS USSR, 1986. - 116 p.
Remenets, A. I. Experience in studying the parameters of bottom deformation and intensity of the longshore sediment flow in Terpeniya Bay / A. I. Remenets // Problems of geography of the Pacific Ocean and near-Pacific regions. Vladivostok. - 1975. - P. 54-63
Remenets, A. I. Evolution of the lagoon coast of Piltun Bay in the Holocene / A. I. Remenets, I. N. Klimov // Development of the natural environment in the Pleistocene. - Vladivostok: Far Eastern Scientific Centre of the Academy of Sciences of the USSR, 1981. – P. 123-125.
Kononov, Yu. I. On the development of abrasion in the Amur estuary / Yu. I. Kononov, V. A. Morin // Climatic geomorphology of the Far East. - Vladivostok, 1975a. - P. 56-59.
Kononov, Yu. I. The role of wind sand transport in the formation of the relief of the north-western part of Sakhalin / Yu. I. Kononov // Proceedings of the X1V scientific conference of the Far East State University. - Vladivostok, 1969. - P. 47-51.
Kononov, Yu. I. Conditions and character of clastic material inflow into the Amur estuary / Yu. I. Kononov, N. N. Kononova // Problems of geography and geomorphology of the Soviet Far East. - Vladivostok, 1992. - P. 154-163.

Kononov, Yu.I. The role of ice in the formation of the Nevelskii Strait sediments / Yu.I. Kononov, V.A. Morin, N.S. Petrik // Problems of geography of the Pacific Ocean and near-Pacific areas. - Vladivostok, 1975b. - P. 64-67.
Ignatov, E. I. Coastal morphosystem as an object of geomorphological research / E. I. Ignatov // Geomorphology. - 2005. - №. 2. - P. 3-11.
Ignatov, E. I. Geomorphology of bay shores and underwater slope of the Primorsky coast of the Sea of Japan / E. I. Ignatov, M. Y. Lokhin, A. V. Nikiforov, V. V. Frol. - Smolensk: Majesta, 2004. - 192 p.
Ignatov, E. I. Dynamics of sediments in the Zerkalnaya Bay of the Sea of Japan according to the data of mineralogical analysis / E. I. Ignatov, I. S. Prokhodsky, M. Turkova // Vopr. of solid matter differentiation on the continental shelf. - Moscow: Nauka. - 1976. - P. 61-65.
Ignatov, E. I. Determination of lithodynamic elements by the LMP method on the coast of the Sea of Japan in Rudnaya and Zerkalnaya Bays / E. I. Ignatov, D. Katagoshchin, I. S. Prokhodsky & Robsman // Oceanology. - 1979. - T. 19. - №. 1. - P. 181-185.
Ignatov, E. I. Evolution of coastal morphosystems / E. I. Ignatov // Geomorphology at the turn of the XXI century. IY Shchukin Readings. Proceedings. - Moscow: Iz-vo Geographical faculty of Moscow State University. - 2000. - P. 149-153.
Hayes, M. O. Geomorphology of the southern coast of Alaska / M. O. Hayes, C. H. Ruby, M. F. Stephen, & S. J. Wilson // Coastal Engineering. – 1977. – P. 1992–2006.
Molnia, B. F. Rapid Shoreline Erosion And Retreat At Icy Bay, Alaska A-Staging Area For Offshore Petroleum Development / Molnia, Bruce F. // Offshore Technology Conference. – 1977. – P. 115–125.
Jordan, J. W. A 5000 year record of intertidal peat stratigraphy and sea level change from northwest Alaska / J. W. Jordan, O. K. Mason // Quaternary International. – 1999. – T. 60. – №. 1. – P. 37–47.
Mason, O. K. Chronology and paleoclimate of storm-induced erosion and episodic dune growth across Cape Espenberg Spit, Alaska, USA / O. K. Mason, D. M. Hopkins, L. Plug // Journal of Coastal Research. – 1997. – P. 770–797.
Lander, J. F. Tsunamis Affecting Alaska 1737–1996/ J. F. Lander // No. 31. US Department of Commerce, National Oceanic and Atmospheric Administration, National Environmental Satellite, Data, and Information Service, National Geophysical Data Center, 1995. – 195 p.
Boothroyd, J. C. Processes, bar morphology, and sedimentary structures on braided outwash fans, northeastern / J. C. Boothroyd, G. M. Ashley // Gulf of Alaska: Society of Economic Paleontologists and Mineralogists. – Special Publication no. 23. – 1975. – P. 193–222.
Gorin, S. L. Modern hydrological regime of the lower Penzhina River and first information about hydrological processes in its estuary (based on the results of the expedition in 2014) / S. L. Gorin, M. V. Koval, A. A. Sazonov, P. N. Tersky // Studies of aquatic biological resources of Kamchatka and the North-West Pacific Ocean. - 2015. - №. 37.
Gorin, S. L. Estuaries of the Kamchatka Peninsula: theoretical approaches to the study and hydrological and morphological typing. Results of 10 years of research / S. L. Gorin // Studies of aquatic biological resources of Kamchatka and the North-West Pacific Ocean. - 2012. - №. 27. - P. 5-12.
Korotkiy, A. M. Anomalous natural processes and their impact on the state of geosystems in the south of the Russian Far East / A. M. Korotkiy, V. V. Korobov, G. P. Skrylnik. - Vladivostok: Dalnauka, 2011. - 265 p.
Skrylnik, G. P. Leading factors of the Far East relief development and its place in the general climomorphogenesis of the Northern Hemisphere continents / G. P. Skrylnik // Studies of global factors of the Far East climomorphogenesis. - Vladivostok: Far Eastern Scientific Centre of the Academy of Sciences of the USSR, 1979. – P. 7-31.
McLaren, P. Longshore transport variability of beach face grain size: Implications for dune evolution / P. McLaren Discussion of: C. Hallin, B. Almström, M. Larson, and H. Hanson // Journal of Coastal Research. – 35 (4). – 751–764 Journal of Coastal Research. – 2019. – T. 35. – №. 5. – P. 1362–1367.

McLaren, P. The environmental implications of sediment transport in the waters of Prince Rupert, British Columbia, Canada: a comparison between kinematic and dynamic approaches / P. McLaren // Journal of Coastal Research. – 2015. – T. 32. – №. 3. – P. 465–482.
Sunamura, T. Rocky coast processes: with special reference to the recession of soft rock cliffs / T. Sunamura, // Proceedings of the Japan Academy. – 2015. – Series B, 91(9). – P. 481–500.
Mason, O. K. Narratives of shoreline erosion and protection at Shishmaref, Alaska: The anecdotal and the analytical / O. K. Mason, J.W. Jordan, L. Lestak, W. F. Manley // Pitfalls of Shoreline Stabilization. – Dordrecht: Springer, 2012. – P. 73–92.
Jones, B. M. Increase in the rate and uniformity of coastline erosion in Arctic Alaska / B. M. Jones, C. D. Arp, M. T. Jorgenson, K. M. Hinkel, J. A. Schmutz, P. L. Flint // Geophysical Research Letters. – 2009. – 36(3). – L03503.
Gorokhovich, Y. Historical and future coastal changes in Northwest Alaska / Y. Gorokhovich, A. Leiserowiz // J. of Coastal Research. – 2011. – 28(1A). – P. 174–185.
Gibbs, A. E. National assessment of shoreline change – A GIS compilation of vector shorelines and associated shoreline change data for the north coast of Alaska U. S.-Canadian border to Icy Cape / A. E. Gibbs, K. A. Ohman, B. M. Richmond // U. S. Geological Survey Open-File Report 2015a/ – p. 1030. (https://doi.org/10.3133/ofr20151030.)
Gibbs, A. E. National assessment of shoreline change – Historical shoreline change along the north coast of Alaska, U. S. [Text] / A. E. Gibbs, B. M. Richmond; Canadian border to Icy Cape: U. S. Geological Survey, 2015b. – №. 2015–1046. – 96 p.
Gibbs, A. E., Richmond B. M. National assessment of shoreline change – Summary statistics for updated vector shorelines and associated shoreline change data for the north coast of Alaska, U. S. [Text] / A. E. Gibbs, B. M. Richmond; Canadian border to Icy Cape. U. S. Geological Survey Open-File Report 2017. – 21 p.
Uda, T. Japan's beach erosion: reality and future measures. Advanced Series on Ocean Engineering Vol 31 / T. Uda. – Singapore: World Scientific, 2010. – 418p.
Afanas'ev V. V. V. Morpholithodynamic processes and coastal development of the contact zone of the subarctic and temperate seas of the North Pacific // Yuzhno-Sakhalinsk: IMGiG FEB RAS. - 2020. 234 p.
Leontiev, I. O. Dynamics of the lagoonal shore of northeastern Sakhalin on the example of the system of Nyisky Bay and Plastun Spit / I. O. O. Leontiev, V. V. Afanas'ev // Oceanology. - 2015. Vol. 55. № 4. - P. 618-625. – (https://doi.org/10.7868/S0030157416030138).
Longinov, V. V. From the history of studying the dynamics of the coastal zone. Issues of the study of marine shores / V. V. Longinov, Ed: V. P. Zenkovich. P. Zenkovich. - Moscow: Academy of Sciences Publishing House, 1959. - P. 161-197.
Zenkovich, V. P. Fundamentals of the doctrine of the sea shore development [Text] / V. P. Zenkovich. - M.: Izd-vo of the USSR Academy of Sciences, 1962. - 710 p.
Leontiev, O. K. Geomorphology of sea shores and bottom / O. K. Leontiev. - M., Izd-vo MSU, 1955. - 376 p.
Longinov, V. V. Dynamics of the coastal zone of the non-tidal seas / V. V. Longinov. - Moscow: Izd-constituion of the USSR Academy of Sciences, 1963. - 379 p.
Budanov, V. I. Methods of expeditionary research in the coastal zone of the sea / V. I. Budanov. - Moscow: Nauka, 1964. - 227 p.
Kaplin, P. A. The Modern History of the World Ocean / P. A. Kaplin. - Moscow: Izd vo MSU, 1973. - 265 p.
Aibulatov, N. A. Investigation of the longshore movement of sandy sediments in the sea / N. A. Aibulatov. - M. Nauka, 1965. - 160 p.
Dolotov, Y. S. Sediment differentiation and layering of coastal sediments [Text] / Y. S. Dolotov, R. B. Zharomskis, V. I. Kirlis. - Moscow: Nauka, 1982. - 184 c.
Safyanov, G. A. Coastal zone of the ocean in the XX century [Text] / G. A. Safyanov. - Moscow: Mysl, 1976. - 264 p.
Safyanov, G. A. Dynamics of the coastal zone of the seas / G. A. Safyanov. - Moscow: Moscow State University Publishing House. - 1973. - 174 p.

Skrylnik, G. P. Anomalous natural processes and phenomena of the Russian Far East / G. P. Skrylnik // Uspekhi sovremennoi naukhestvennosti. - 2016. - № 10. - P. 114-124, (http://natural sciences.ru/en/article/view?id=36892 date of address: 07. 11. 2019).
Nikolskaya, V. V. The peculiarity of the relief of the megaboard of the Sea of Okhotsk / V. V. Nikolskaya, G. P. Skrylnik // Climatic geomorphology of the Far East. - Vladivostok: TIG DVNTs of the USSR Academy of Sciences. - 1975. - P. 44-49.
Nikolskaya, V. V. On natural tendencies of development of physiographic provinces of the south of the Far East / V. V. Nikolskaya.- Novosibirsk: Nauka, 1974. - 127 p.
Skrylnik, G. P. The role of continentality and oceanicity in the sustainable development of the south of the Far East / G. P. Skrylnik // Regions of new development: development strategy. (Proceedings of the International Scientific Conference. Khabarovsk, 15-17 September 2004). - Khabarovsk: Khabarovsk Krai Government, Institute of Economic Policy of the Far East Branch of the Russian Academy of Sciences, Priamurskoye GO. - 2004. - P. 230-234.
Likhacheva, E. A. Physical essence of geomorphological boundaries / E. A. Likhacheva, A. N. Makkaveev, G. P. Lokshin // Geomorphology. - 2010. - № 2. - P. 3–10.(https://doi.org/10.15356/0435-4281-2010-2-3-10)
Baklanov, P. Ya. Geographical and geopolitical factors in regional development / P. Ya. Baklanov // Regional Studies - 2014. - № 2. - P. 4-10.
Likhacheva, E. A. Analysis of geomorphological systems: basic concepts / E. A. Likhacheva, D. A. Timofeev // Geomorphology. - 2006. - №2. - P. 14-21.
Likhacheva, E. A. To the hierarchy of geomorphological systems from the position of their organisation / E. A. Likhacheva, D. A. Timofeev // Geomorphology. - 2007. - №4. - P. 3-6. (https://doi.org/10.15356/0435-4281-2007-4-3-8)
Dzerdzeyevksii, B. L. General circulation of the atmosphere and climate [Text] / B. L. Dzerdzeyevksii - M. Nauka, 1975. - 288 p.
Tunegolovets, V. P. Climate and hydrography / V. P. Tunegolovets, B. I. Gartzman, V. V. Krokhin // Geosystems of the Russian Far East at the turn of XX-XXI centuries. - 2006. - P. 119-143.
Lisogurskii, N. I. Distribution of the monsoon over East Asia and the degree of its stability / N. I. Lisogurskii, A. Z. Petrichev // Meteorology and Hydrology. - 1980. - VOL. 5. - P. 54-59.
Skrylnik, G. P. Leading climatic factors of exogenous relief development / G. P. Skrylnik // Geosystems of the Russian Far East at the turn of XX-XXI centuries. - 2006. - P. 77-81.
Alekseev, B. A. Global model of modern landscapes of the world / B. A. Alekseev, G. N. Golubev // Functioning and current state of landscapes / Edited by K. N. Diakonov, E. P. Romanova. - Moscow: Izd. Dom "Gorodets", 2004. - P. 319-325.
Dyakonov, K. N. Influence of the oceans on the physiographic zonality according to satellite data on photosynthesis / K. N. Dyakonov, D. L. Varlygin, A. Yu Reteyum // Bulletin of Moscow University. - Series 5: Geography. - 2017. - № 2. - P. 11-15.
Ermakov, V. A. New understanding of geodynamic relations of the Commander Basin and the Aleutian Island Arc (in modern times) / V. A. Ermakov, A. V. Ermakov // Physico-chemical and petrophysical studies in Earth Sciences. - 2016. - P. 108-115.
Bark, K. Aleutian island arc and the continental margin of Alaska / K. Bark // Continental margins and island arcs. - M., Mir. - 1970. - P. 178-184.
Kolomyts, E. G. Bioclimatic system of the Kuril Islands / E. G. Kolomyts // Izvestia of the Russian Academy of Sciences. Series geographical. - 2016. - №. 3. - P. 79-90.
Krasny L. I. Geological structure of the northwestern part of the Pacific mobile belt. - L. I. Krasny (ed.). - Nedra, 1965. - 515 p.
Hanchuk, A. I. Tectonic scheme of the Russian Far East / A. I. Hanchuk, V. V. Golozubov, N. A. Goryachev, L. I. Popeko. - DVGI DVO RAN, 2009.
Matul, A. G. Quaternary biostratigraphy and palaeoceanology of the Sea of Okhotsk and other subarctic regions / A. G. Matul. - M: GEOS, 2009. - 182 p.
Ustinova, E. I. Interannual variability of thermal conditions of the Sea of Okhotsk / E. I. Ustinova, Y. D. Sorokin, G. V. Hen // Izvestia TINRO. - 2002. T. 130. - №. 1-1. - P. 44-51.

Hen, G. V. Main regularities of multi-year changes in the ice cover of the Sea of Okhotsk and the Bering Sea / G. V. Hen // Integrated studies of the Okhotsk Sea ecosystem. - M.: VNIRO, 1997. - P. 64-67.
Marine Atlas. Vol. 2. [Text] / M., Ministry of Defence of the USSR; Navy, 1953. - 82 c.
Nikitenko, Yu. P. Map of modern vertical movements of the earth crust of Sakhalin Island. Scale 1 : 1 250 000 / Yu. P. Nikitenko. - M.: GUGK, 1981.
Zemtsova, A. I. Climate of Sakhalin [Text] / A. I. Zemtsova. - L.: Gidrometeoizdat, 1966. - 197 p.
Aleksandrov, S. M. Sakhalin Island [Text] / S. M. Aleksandrov. - Moscow: Nauka, 1973. - 182 p.
Petrov, A. G. Regionalisation of the Tatar Strait water area by the character of ice cover formation in severe ice years / A. G. Petrov // Hydrometeorological and ecological conditions of the Far East seas: assessment of the impact on the marine environment: Thematic issue of DVNIGMI No. 2. - Vladivostok: Dalnauka, 1999. - P. 138-145.
Skokov, R. M. Distribution and stability of landfast ice off the coast of Sakhalin: Cand. Sci. Geogr. Sciences / R. M. Skokov // Л.: 1982. - 158 p.
Tambovsky, V. S. Characteristics of ice drift velocity off the north-east coast of Sakhalin Island under the influence of tides and wind / V. S. Tambovsky, G. V. Shevchenko // Proceedings of DVNIGMI Thematic Issue No. 2. - Vladivostok: Dalnauka, 1999. - P. 114-137.
Archikov, E. I. The role of ice formations in the development of coastal geosystems of the Sea of Okhotsk / E. I. Archikov, L. E. Stepanova, I. S. Mayorov. - Vladivostok : Izd. of Far Eastern University, 1989. - 109 p.
Atlas of the Sakhalin Region / Moscow, 1967.
Tambovsky, V.S. Morphometry and ice dynamics in the water area of the north-eastern shelf of Sakhalin Island / V.S. Tambovsky, V.F. Putov, G.V. Shevchenko, E.A. Tikhonchuk // Nature protection, monitoring and development of the Sakhalin shelf. - Yuzhno-Sakhalinsk, 2001. - P. 123-142.
Shevchenko, G. V. Dynamics of ice drift on the north-eastern shelf of Sakhalin Island according to measurements by radar stations / G. V. Shevchenko, V. S. Tambovsky. - Yuzhno-Sakhalinsk: 2016. - 136 p.
Ponomareva, T. G. Water exchange of the Sea of Okhotsk and the Sea of Japan through the Amur estuary / T. G. Ponomareva // Proc. of DVNIGMI. DVNIGMI. - 1989. - Issue. 36. - P. 123-136.
Ponomareva, T.G. Estimation of water flow on the borders of the Amur River mouth section / T.G. Ponomareva // Proc. of DVNII. DVNII. - 1982. - Issue. 105. - P. 60-62.
Kato, E. Calculation of extreme values of the runup and surge sea level fluctuations on the south-east coast of Sakhalin Island / E. Kato, Y. V. Lyubitsky, G. V. Shevchenko // Sea level fluctuations in the seas. - Russian Hydrometeorological University. - SPb.: Gidrometeoizdat, 2003. - P. 111-126.
Kato, E. Extreme heights of storm surges on the coast of Sakhalin Island / E. Kato, Y. V. Lyubitsky, G. V. Shevchenko // Shipping and Marine Science. - 2011. - P. 177-193.
Shevchenko, G. V. Manifestations of storm surges in the southern part of Sakhalin Island [Text] / G. V. Shevchenko, Y. V. Lyubitsky, L. N. Kato. - Yuzhno-Sakhalinsk. 1994. - 46 p.
Shevchenko, G. V. Statistical characteristics of storm surges in the southern part of Sakhalin Island / G. V. Shevchenko // Izv. RGO. - 1997. - T. 129. - № 3. - P. 94-107.
Poezhdalova, S. Okhotsk. Variations of the mean level of the Sea of Okhotsk / o-va S. Poezhalova, G. V. Shevchenko // Tsunami and related phenomena. - Yuzhno-Sakhalinsk, 1997. - P. 131-144.
S. Sedaeva, S. S. On the relationship between seasonal variations of sea level and atmospheric pressure in the area of the Kuril Ridge / S. S. S.. Sedayeva, G. V. Shevchenko // Dynamic processes on the shelf of Sakhalin and Kuril Islands. - Yuzhno-Sakhalinsk, 2001. - P. 81-93.
Atlas of Pacific Ocean waves and winds / L., Gidrometeoizdat, 1966. - 100 p.
Atlas of waves and winds of the Sea of Japan / (compilers: V. I. Perfilieva, A. A. Petroshenko, A. M. Polyakova). - Main Directorate of Hydrometeorological Service under the USSR Council of Ministers. Primorsky Administration of Hydrometeorological Service. - Vladivostok, 1963. - 120 p.

Koshinsky, S. D. Regime characteristics of strong winds in the seas of the Soviet Union: the North of the Sea of Japan, the Sea of Okhotsk and the Bering Sea / S. D. Koshinsky - L. : Gidrometeoizdat, 1976. - Ch. 2. 230 p.
Popov, B. A. Principles of selection of initial data for calculation of wave energy load / B. A. Popov, V. A. Sovershaev // Coastal zone of the sea. - Moscow: Nauka, 1981. - P. 47-52.
Kovalev, P. D. Renewable energy resources of the Sakhalin region / P. D. Kovalev, D. P. Kovalev, G. V. Shevchenko. - Publisher: Dalnauka 2015. - 216 p.
Afanas'ev V. V. Report on lithodynamic and engineering hydrometeorological surveys for the object 05276 "Development of the North-Veninskoye gas condensate field (SV GCF)" / V. V. Afanas'ev. - LLC "Dalmorberegotekhnika", Yuzhno-Sakhalinsk, 2015a. - 115 p.
Afanas'ev V. V. Report on the research work "Study of morpholithodynamics of the banks of the Odoptu Sea to justify protection from scour of production sites and transport corridors / V. V. Afanas'ev. - RN-Sakhalinmorneftegaz LLC, Dalmorberegotechnika LLC, Yuzhno-Sakhalinsk, 2015b. - 106 p.
Afanas'ev, V. V. Report on the research work "Investigation of morpholithodynamics of the banks of the Odoptu Sea to justify the protection from scour of production sites and transport corridors of LLC RN-Sakhalinmorneftegaz" / V. V. Afanas'ev. - Yuzhno-Sakhalinsk, 2015c. – 107 p.
Shevchenko, G. V. Assessment of extreme sea level fluctuations in the area of the Tym River estuary / G. V. Shevchenko, A. V. Fine, A. B. Rabinovich et al. // Natural catastrophes and natural disasters in the Far East region. - Vladivostok, 1990. - T. 1. - P. 253-275.
Dunayev, N. N. Modern dynamics of the accumulative bank composed by pyroclastics of the underwater volcanic eruption / N. N. Dunayev // Geodynamic processes and natural disasters. - 2019. - P. 137-137.
Kaplunenko, D. D. Typical fields of wind waves and quicksand in the water area of the North Pacific Ocean / D. D. Kaplunenko, A. M. Polyakova, S. S. Marchenko // Oceanology. - 2011. - T. 51. - №. 5. - P. 788-795.
Polyakova, A. M. Atmospheric circulation and wind wave and quicksand fields in the North Pacific Ocean / A. M. Polyakova, D. D. Kaplunenko // Far East Seas of Russia. - 2007. - P. 112-135.
Polyakova, A. M. Wind waves and ripples in the North Pacific Ocean. / A. M. Polyakova, S. S. Sugak // Meteorology and Hydrology. - M., Publishing Centre "Meteorology and Hydrology", 2004. - № 2. - P. 79-87.
Vykhovanets, G. V. Aeolian process on the sea shore / G. V. Vykhovanets. - Odessa: Astroprint, 2003. - 368 p.
Afanas'ev, V. V. Evolution of the Far Eastern Seas Coast in the Holocene / V. V. Afanas'ev // Coastal Evolution in the Conditions of Ocean Level Rise. - M.: Izd-vo MSU, 1992. - P. 166-174.
Korotky, A. M. Aeolian relief of Primorye and adjacent territories of East Asia (palaeogeographical aspect) / A. M. M. Korotkiy // Geomorphology. - 2007. - №. 4. - P. 79-95.
Korotkiy, A. M. Coastal dunes-indicator of global cooling (Kunashir Island, Kuril Islands) / A. M. Korotkiy, N. G. Razhigayeva, L. M. Mokhova, L. A. Ganzey, T. A. Grebennikova, V. B. Bazarova // Pacific Geology. - 1995a. - T. 15. - №. 1. - P. 53-59.
Badyukova, E. N. Relief of seaside dunes as an indicator of sea level fluctuations / E. N. Badyukova, G. D. Solovyova // MSU Vestn. MSU. - 1997. - №. 5. - P. 10-5.
Badyukova, E. N. Coastal aeolian forms and sea level fluctuations / E. N. Badyukova, G. D. Solovyova // Oceanology. - 2015. - T. 55. - №. 1. - P. 139-139.
Aagaard, T. Environmental controls on coastal duneformation; Skallingen Spit, Denmark / T. Aagaard, J. D. Orford, A. S. Murray //Geomorphology. Aagaard, J. D. Orford, A. S. Murray //Geomorphology. - 2007. - 83. - P. 29-47
Porter, S. C. Late Pleistocene eolian sediments related to pyroclastic eruptions of Mauna Kea Volcano, Hawaii / S. C. C. Porter // Quaternary Research. - 1997. - T. 47. - №. 3. - P. 261-275.
Botvinkina, L. N. Genetic types of sediments of active volcanism. / L. N. Botvinkina. - Moscow: Nauka, 1974. 318 p.

Melekestsev, I. V. Novel terrestrial and submarine volcanism of the Kuril island arc / I. V. Melekestsev // Novel and modern volcanism on the territory of Russia Ed. by N. P. Laverov - M.: Nauka, 2005. - P. 233-335.
Afanas'ev, V. V. Vetrovoy isthmus of iturup island-holocene strait / V. V. V. Afanas'ev // IOP Conference Series: Earth and Environmental Science. - IOP Publishing, 2019. - T. 324. - №. 1. - P. 012029.
Korotkiy, A. M. Coastal dunes-indicator of global cooling (Kunashir Island, Kuril Islands) / A. M. Korotkiy, N. G. Razhigayeva, L. M. Mokhova, L. A. Ganzey, T. A. Grebennikova, V. B. Bazarova // Pacific Geology. - 1995b. - T. 15. - №. 1. - P. 53-59.
Razhigayeva, N. G. Sedimentation settings of island territories in the Pleistocene-Holocene / N. G. Razhigayeva, L. A. Ganzey. - Dalnauka Publishing House, 2005. - 365 p.

Morphology and Dynamics of the Shores of the Contact Zone of Temperate and Subarctic Seas

V. Afanas'ev Viktor and E. I. Ignatov

Abstract Detailed studies of morphology and dynamics of the coasts of the contact zone of temperate and subarctic seas are considered on the example of Sakhalin Island-the contact zone between the subarctic Sea of Okhotsk and temperate Sea of Japan. The modern geomorphological structure of the Sakhalin coast is the result of a long and complex interaction of endogenous and exogenous processes of relief formation. This has determined the use of a wide range of methods to analyse the geomorphological structure of the tyerritor.

Keywords Contact areas of the seas · Sakhalin Island · Coastal morphodynamics

1 The Coast of the Coastal-Marine Plain of North-West Sakhalin Island

The northwestern coast of Sakhalin lies within a distinctive morphostructural zone where the Sikhote-Alin and Hokkaido-Sakhalin folded regions converge. Remote sensing and morphostructural interpretation of satellite imagery have revealed a system of overlapping semicircular formations along western Sakhalin that extend to the coastline. Similar semicircular structures have also been documented along the northern coast of Sikhote-Alin.

These large-scale semicircular morphostructures are primarily composed of intrusive, volcanic, and volcanic-sedimentary formations associated with the East Sikhote-Alin volcanic belt. They incorporate Cretaceous volcanic-sedimentary complexes that resemble those found in central-type morphostructures (CTM) of the West Sakhalin series.

V. Afanas'ev Viktor (✉)
Sakhalin State University, Yuzhno-Sakhalinsk, Russia

E. I. Ignatov
Faculty of Geography, Moscow State University, Moscow, Russia

T. Chaplina (ed.), *Coastal Geosystems of Cold Seas Under Conditions of Climatic and Anthropogenic Changes*, Earth and Environmental Sciences Library, https://doi.org/10.1007/978-3-032-12247-6_2

The correspondence in outline between the West Sakhalin and Sikhote-Alin semicircular structures, along with their shared Upper Mesozoic conformational complexes, suggests they form a unified system of CTMs. Their formation is likely the result of crustal fragmentation at the continental margin, involving the emplacement of large plutons and thrust faulting. From this perspective, the Lazarevskoye and Pogibinskoye "half-ring" formations in northwestern Sakhalin can be interpreted as segments of a single CTM approximately 140 km in diameter (Fig. 1).

One of the main elements of this structure is a large magmatic formation of basic composition intruded into the sedimentary volcanogenic strata (Kononov 1986).

The arc-shaped depression framing the uplift in the east has a higher hypsometric position in the present-day relief due to the surface skew as a result of the Jolokor-Engizpal Pliocene mountain uplift and the formation of outflow cones within its boundaries. The depression is well distinguished on the CS, and the Tengin, Vagis and Tyk synclinal zones can be traced as a single formation. The structure of the sedimentary strata composing the depression indicates the inherited from the Palaeogene deflection relative to the central parts of the structure. The thickness of Cenozoic sediments varies from 3500 m on the northern flank of the zone to 8000 m on the southern flank (Mishakov 1983).

The absence of regional angular unconformities and dislocations in the Quaternary sediments indicates that the formation of the structural plan, which began in the Late Miocene, continued in the Pliocene and Quaternary and has not ended at present (Khudyakov 1977).

The influence of Holocene and Upper Pleistocene tectonic movements on coastal formation—expressed through the morphometric and morphodynamic characteristics of the coastal zone—is now widely accepted. In northwestern Sakhalin, integrated analysis of the coastal-marine plain and its underlying morphostructure reveals that under conditions of dominant longshore sediment transport, the largest Holocene accumulative formations—Tyk Spit and a system of Azov-type spits—are situated where marginal-depressional elements of the morphostructural framework intersect the coast. These areas have experienced the highest sedimentation rates throughout the Neogene-Quaternary period.

The sediment volumes comprising these features are substantial: approximately 152 million m^3 for Tyk Spit and 96 million m^3 for the spit system. It is important to note that the true scale of accretion is only apparent when considering their submarine portions, which contain about 90% of the total sediment mass.

Further evidence of the inherited subsidence of these depressional zones during the Holocene is provided by a distinctive ridge-lake complex on the northern flank of the Tyksko-Vagysskaya Depression. This system aligns with the principal structural trends of the Lazarevsko-Pogibinskaya interstructural trough complex (ITC). Analogous ridge-lake systems in Western Siberia have been interpreted as resulting from the displacement of lacustrine-marsh formations in response to tectonic tilting.

The relief structure of the alluvial-marine plain of north-western Sakhalin is generally rather poorly studied (Svitoch 2003). This is most clearly expressed by divergent views on the problem of cyclic geomorphological levels. One of the most comprehensive summaries of Sakhalin geomorphology describes 4 cyclic levels

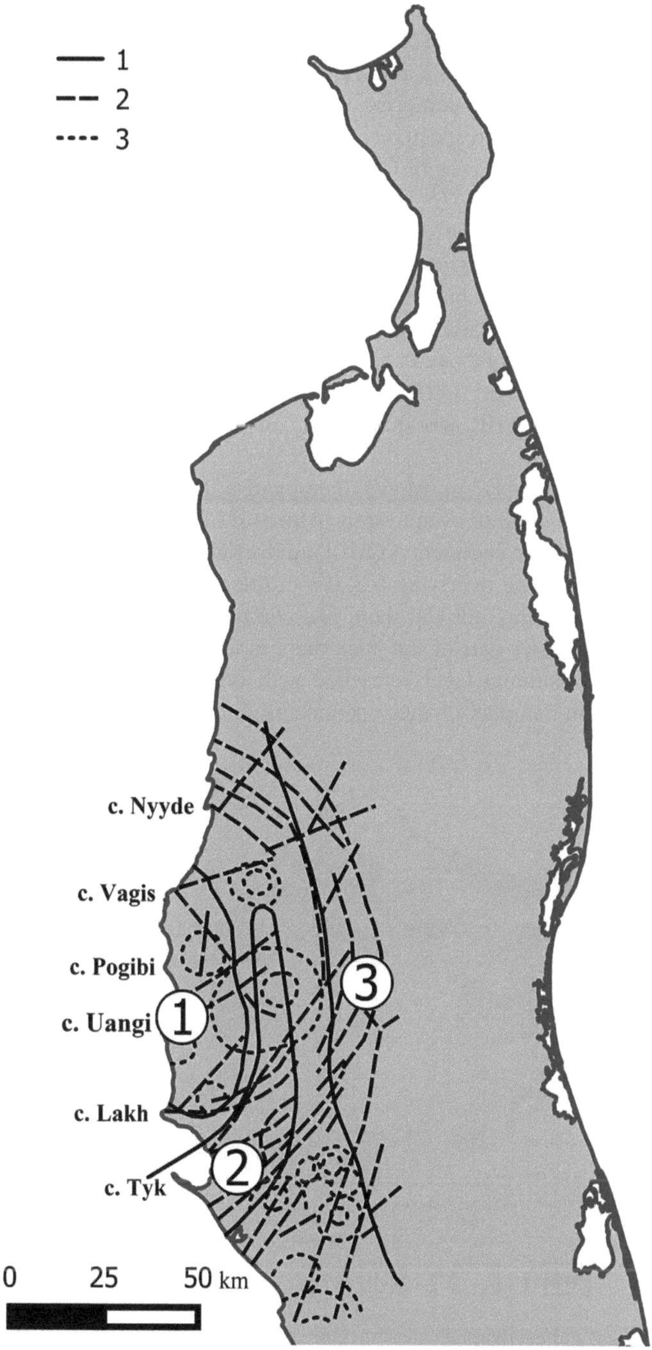

Fig. 1 Sakhalin segment of the Lazarevsko-Pogibinskaya morphostructure of the central type. 1—boundaries of structural facies zones; 2—framework (disjunctive) elements of the Lazarevsko-Pogibinskaya CTM; 3—brachiform anticlinal folds of cladding. Figures in circles: 1—Pogiba magmatic formation of basic composition, intruded into the sedimentary volcanogenic strata; 2—Vagyssko-Tyk depression; 3—Jolokor-Engizpal Pliocene mountainous uplift

(Solovyev 1971). Two cyclic levels are described in the work of A. N. Aleksandrova (Alexandrova 1987). The low level is represented by the 3–8 m surface (Q $III_{2\text{-}4}$.) of the north-west coast of Sakhalin and the 12–15 m high terraces (Q $III_{3\text{-}4}$) south of Cape Tyk. Deposits Q $III_{1\text{-}2}$ comprise a terrace level up to 40 m high. The lower horizon of $QIII_3$ and $QIII_1$, respectively, of both terraces are marine terrigenous sediments, while the upper horizon of $QIII_4$ and $QIII_2$ are regressive marine and continental sediments.

A detailed analysis of the geological and geomorphological structure of the alluvial-marine plain was carried out in several key areas. Along with spore-pollen, diatom and granulometric characteristics of sediments, more than 40 radiocarbon age determinations were made here.

In the Nevelsky Strait area, Late Pleistocene sediments 38,340 ± 1020 (FESU-112a); 28,800 ± 1425 (FESU-112); 30,087 ± 2400 (FESU-111); 25,223 ± 434 (FESU-110) are ubiquitously exposed at the base of the section (Fig. 2).

At elevations of 0.5–1.7 m, M. V. Cherepanova determined a diatom complex characteristic of a lagoon or even a strait (Afanas'ev 1992).

The lagoonal-marine sediments ($QIII_3$) are overlain by a thin algal peat layer, in some areas by two. The overlying 0.3–0.4 m thick interbed is younger by about 10,000 years. Above this horizon there is a complexly built layer of lake-alluvial sediments. In the lower part of the pack there is a layer of grey sandy loam with traces of cryodepositional layers enriched with organics. Unfortunately, diatoms were not found in samples of these sediments. Sands, loams, gravel and pebble

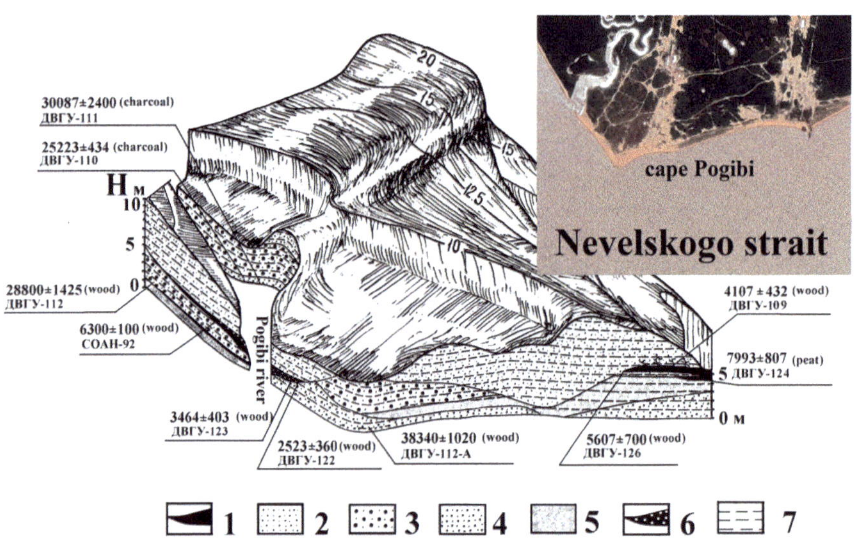

Fig. 2 Geological and geomorphological structure of the alluvial-marine terrace (Cape Pogibi). 1—peat, 2—fine-medium grained aeolian sands, 3—tin pebbles with gravel and sandy aggregate (alluvium), 4—silty fine-grained lagoonal sands, 5—sandy-pebble wave-surf deposits, 6—peat with silty-pelitic aggregate. Baltic height system (1977)

sediments characterised by bog, lake, alluvial and transitional diatom complexes occur even higher up.

These sediments also contain redeposited remains of algae of ancient species. The section of the coastal-marine plain sediments is crowned by Holocene sediments represented by peat bogs, eolian sands and estuarine-lagoonal facies embedded in the Upper Pleistocene strata. Peat accumulation, which continues to the present day, is a significant factor in relief formation. Under relative homogeneity of conditions, the higher the hypsometric level of the site, the later the peat bog starts to form on it. Thus, one of the geomorphological effects of peat accumulation—smoothing of relief irregularities—is manifested. Aeolian processes can also fulfil a similar function, however, they mainly play the opposite role on the coast. In the area of the mouth of the Pogibi River, aeolian forms reach +25 metres. The age of wood remains buried here at the base of the dunes is 4107 ± 432 years (FESU-109).

One of the most interesting features of the relief of the coastal plain of northwestern Sakhalin studied in the area of Nevelsky Strait are sand ridges. These formations are up to 10–15 metres high and are very contrasting on the waterlogged surface of the plain. They are confined, except for the Tyk River and the Tyk River. With the exception of the Tyk and Viahtu rivers, they are confined to the left bank of the plain. With the exception of the Tyk and Viahtu rivers, they are confined to the left bank of the rivers, which is consistent with the present-day resultant longshore drift. Let us consider two sections of the sediments composing the ridges in the area of Cape Vagis and Cape Uanga.

The following section is described in the area of Cape Vagis (Depth in metres):

0.00–0.30 soil
0.30–3.00 dark grey fine-medium sand with thin interlayers of rudimentary soils
3.00–3.30 light grey dusty sand with interlayers enriched with organic remains
3.30–3.90 brown tin-grey fine-medium grained sand with interlayers of rudimentary soils
3.90–4.90 interbedded grey fine-medium grained sand with organic enriched layers
4.90–5.50 brown fine-medium-grained sand with organic enriched layers, cryogenic metamorphosis (microcline) observed
5.50–5.20 brown multigrained sand with interlayers of organic remains
5.20–5.50 to groundwater table grey fine-medium sands

The character of layering, the presence of scattered organic matter and thin interlayers of rudimentary soils, the granulometric spectrum similar to modern aeolian sediments allows us to make an assumption about the predominance of the aeolian factor in the formation of the mane. Unfortunately, diatoms were not found even in enriched samples.

The second transect is described in the area of Cape Uanga, 500 metres upstream from the mouth of the river of the same name (Depth in metres):

0.00–0.30 soil
0.30–4.80 multigrained sands with thin interlayers of fine gravel and strongly tiniferous sands.

4.80–5.80 similar to overlying sediments with organic inclusions typical of modern aeolian sediments.
5.80–7.30 multigrained grey sand with interlayers (1–2 cm) of dark grey loam, cryogenically deformed soil layer in the upper part of the horizon
7.30–6.30 multigrained grey sand with interlayers (1–2 cm) of dark grey loams

Unlike the Cape Vagis section, this area contains interlayers of fine gravel. The gravel content increases upward through the section, while the proportion of fine sand decreases correspondingly.

Analysis of diatom flora revealed that sediments between 4.80–6.30 m are characterized by a freshwater diatom complex, alongside some redeposited specimens. At a depth of 1.3 m, where diatoms are less abundant, a mixed assemblage was identified, which also included redeposited forms. The presence of *Paralia sulcata* (a marine species) indicates the proximity of the sea.

Thus, the available information allows us to attribute the forms under consideration to alluvial formations significantly reworked by aeolian processes. The presence of sediments with traces of cryodislocations at the base of the section, the character of relationships with marine, estuarine-lake sediments and Holocene peat bogs allow us to assume their Late Quaternary age ($QIII_{3-4}$). Interestingly, in the Viakhtu area, a radiocarbon date of 26,830 ± 940 (FESU-188) was obtained from sediments of similar composition and morphometric parameters; the presence of peat-gleyey interbeds allowed us to attribute the thickness in question to regressive lake-alluvial sediments (Aleksandrova 1982).

Three geomorphological levels were identified in the area of Viakhtu settlement, located on the bank of the lagoon—estuary of the river of the same name (Fig. 3).

The first level is presented by modern aeolian surf deposits with absolute marks of the rear edge of the beach up to 5 m and the modern marine terrace, which has two levels here—"sisigian", flooded as a result of high tides and wind surges, and "quadrature", flooded almost daily. The surface of the high lyde is well marked by runoff channels. The age of the lowland peat bog overlying black-blue silts is 430 ± 82 (FESU-178). Aeolian sediment accumulation occurs in the section of the laida adjacent to the estuary bar.

Middle Holocene sediments comprise the next level. Peat accumulation on the terrace with an absolute height of 2.5–4 m started about 3500 years ago. Radiocarbon dating of the peatland footwall is 3450 ± 277 (FESU-177) and 3450 ± 390 (FESU-174).

The terrace with absolute elevations of 5–5.5 m began to form as a level of coastal-marine accumulation about 6000 years ago. This is evidenced by the age of the base of the peat bog with a thickness of about 3 metres—5318 ± 320 (FESU-180).

The next level is represented by the surface with absolute marks of 10–20 metres. The study of sediments exposed in the coastal escarpment allowed us to establish that the five-metre height anomaly in the area adjacent to the root part of the spit is due to the deposition of aeolian sands here. Moreover, the accumulation of the uppermost 2–3 metre layer began simultaneously with the formation of peat bog on the high lide 430 ± 78 (FESU-187). Below the section there is an interlacing of

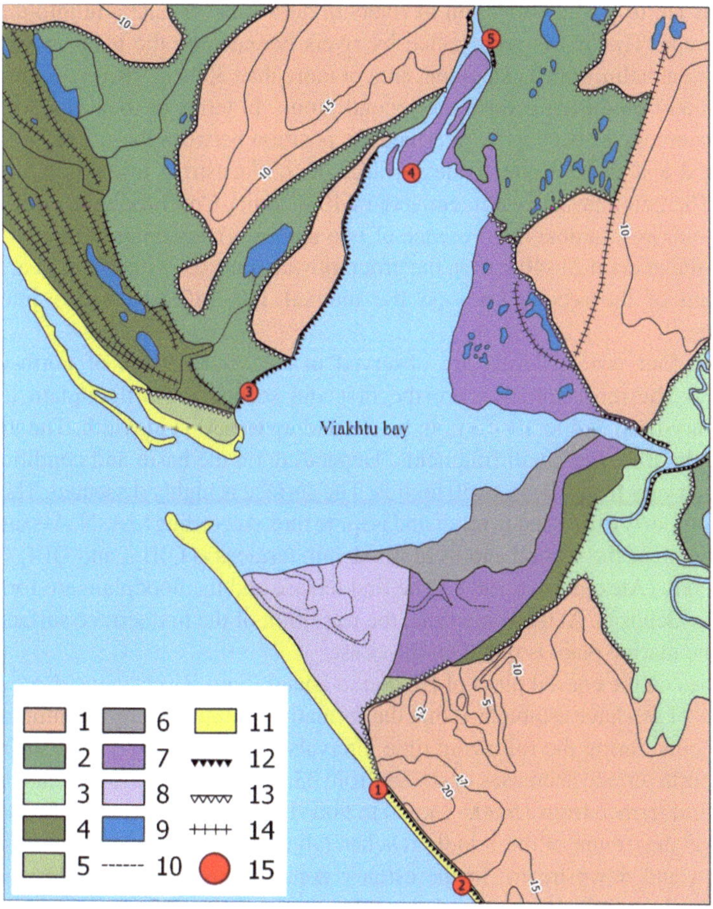

Fig. 3 Geomorphological map of the Viahtu River estuarine zone: 1—Late Quaternary polygenetic accumulative terrace; 2—Middle Holocene lagoonal terrace; 3—Late Holocene lagoonal terrace; 4—Middle-Late Holocene wave and aeolian-surf accumulation surface; 5—Late Holocene wave accumulation surface; 6—tidal accumulative drifts; 7—marshes; 8—marshes exposed to wind-sand flow; 9—residual lakes; 10—runoff channels; 11—modern accumulative forms of the coastal zone; 12—abrasion-erosion ledges of low terraces; 13—dead abrasion-erosion ledges; 14—ancient coastal ramparts and ridges; 15—C^{14} dated geological sections

multigrained sands, loams, gravels and pebbles with peat lenses up to 1 m thick. Age determination of woody remains from the base of one of these peat bogs showed, as noted above, 26,830 ± 940 (FESU-188).

The river valley network in northern Sakhalin exhibits a distinctive centrifugal (outward-flowing) pattern and an unusually positioned main watershed. This anomaly arises because the highest points of the older paleoscarp uplifts, which form the orohydrographic centers, are actually lower than the surrounding younger uplifts that border them (Aleksandrov 1973). Furthermore, an analysis of catchment basin sizes has led to the classification of rivers in northwestern Sakhalin into three distinct groups. The first is represented by rivers bounded by the main watershed of northern Sakhalin, with a catchment area of more than 300 km^2. Rivers with a catchment area of 80-180 km^2 form the second group. In terms of their characteristics, these watercourses occupy an intermediate position between the rivers of the main watershed and those draining the surfaces of the lacustrine low Late Quaternary levels. The latter have a catchment area of 10–60 km^2. Considering the distribution as a whole, we can note the presence of two modes. The maximum of the first one falls on the interval 21–40 km^2 in the group of watercourses with small area, and the maximum of the second one—on the interval 121–140 km^2 in the intermediate group.

Up to four terrace levels are observed in the river valleys of north-western Sakhalin. The most common are the first and second supra-floodplain terraces, which vary from 3–6 to 10–12 m in height and up to 600 m in width. The third terrace can be traced in small fragments. Depending on the basin and conditions, the height can vary from 10–12 to 30 metres. The surface is highly dissected. The fourth terrace is poorly expressed in relief and is quite rare. According to A. N. Alexandrova, the alluvium of the second and third floodplain terraces is $QIII_{3\text{-}4}$ and $QIII_{1\text{-}2}$ in age, respectively (Alexandrova 1987). The first terrace and the floodplain are formed by Holocene deposits. At the same time, the transition of the first terrace surface to the Holocene marine plain is noted on the coast.

Studies on the coastal plain of Hokkaido Island (Saru River) using dated marker horizons of ash have established that the formation of topologically continuous surfaces occurs during the following time intervals (Yanagida 1981). The surface age of the fourth terrace decreases upstream from 33,000 to 9000–15,000 years, and that of the third terrace from 15,000-33,000 to 900–15,000 years. For the second terrace the age is maximum in the middle reaches (about 9000 years) and decreases both upstream and downstream. In the estuary zone its age is 5000–6000 years old. Similarly, the age of the youngest, first terrace changes.

The initial dissection of an emerging marine terrace surface commences with marine regression, characterized by the transgressive development of sub-parallel, closely spaced river valleys exhibiting shallow incision. This dissection is subsequently enhanced by regressive erosion.

Under conditions of geological and geomorphological homogeneity on a coastal alluvial-marine plain, dissection density—a key morphometric parameter of river basins—becomes a particularly valuable indicator.

When a marine terrace is first exposed by a receding sea, its surface is initially carved by a series of parallel, shallow river valleys that form close together. This erosion process then continues to advance uphill (regressive erosion), further increasing the dissection of the landscape.

On a uniform coastal plain, measuring the density of this river network (how many river channels there are per unit area) is especially useful for understanding the basin's morphology, as it provides significant insight when other geological factors are consistent (Fig. 4b).

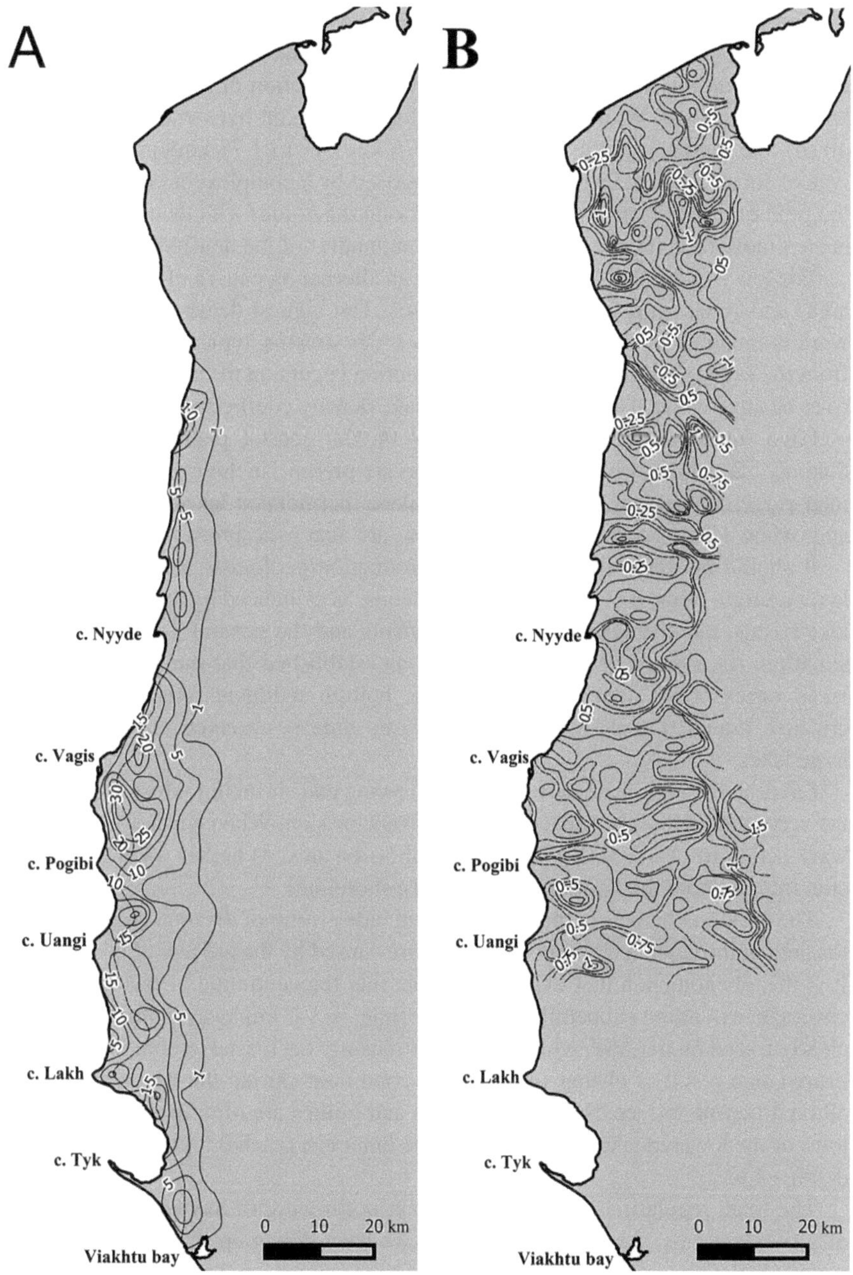

Fig. 4 Cartometric parameters of the coastal lowland: (**a**) map of the territory zooseroseness, (**b**) map of the density of erosion dissection of the territory

The most general patterns of the distribution of the density of dissection in northwestern Sakhalin, which we have identified, include the presence of two zones orientated according to the general, submeridional direction of the coastline and the "main" watershed. Within the first zone, located at high hypsometric levels (over 40 m), the dissection ratio varies from 0.75 km/km^2 to 1.75 km/km^2. The second zone is located at low levels and is characterised by a complex, mosaic structure of the field of values from 0 to 1.0 km/km^2. Along the border with the first zone, there is an almost continuous strip of minimum anomalies of the density of dissection.

The low coastal plain generally features an alternating pattern of highly dissected areas and zones with very few watercourses. The highest density of the river network is found along the watersheds of low-order streams, typically located 5–8 km from the estuary. In contrast, the least dissection occurs on the watersheds of larger river basins, which also show the highest lake density coefficients.

Lake formation is a defining process on the coastal plain of northwestern Sakhalin. Several morphogenetic lake types are present, including: sub-pond lakes, inter-ridge depressions, residual-lagoon lakes, thermokarst lakes, suffusion lakes, and oxbow lakes. So-called "oriented lakes" are also widespread in the region.

It should be noted that the ideas of production geologists about the residual-lagoon origin of most of the lakes are erroneous, as evidenced by the structure of the lake basins, the character of bottom sediments and the general geomorphological situation. As a result of the research, it was established that the depth of coastal lakes varies in the interval of 1.0–2.5 m, bottom sediments are represented by crushed, liquefied peat. A beach up to 10 m wide is observed on the shores of large lakes.

Lakes in the region are currently both growing and shrinking. Most lakes, except for very small ones, are expanding due to bank erosion. When water levels are low, wave action from southern winds creates abrasion arcs. At higher water levels, erosion impacts both the northern and southern shorelines.

The greatest changes in the lake coverage index (ratio of the total area of lakes to the area of the calculated unit of territory) are caused by the process of lake descent. It is this phenomenon that explains the fact that the centreline of the highest lake coverage, extending subparallel to the coastline, is 3–7 km away from it. Lakes are clearly traced on the ASF, which either significantly decreased or completely disappeared as a result of changes in drainage conditions during the destruction of the alluvial-marine terrace. Suffosion hollows and gullies are often formed in the hollows of the lowered lakes. The length of the latter can reach 0.5 km, width −5–6 m, depth −4 m.

The main regularities of lake coverage in north-western Sakhalin, expressed in quantitative terms, are shown on the map we have drawn up (Fig. 4a). Lake areas were calculated using a 1:25,000 scale topo base using a 2 × 2 mm square stick. The elementary unit of area was taken as 18 km^2. Only lakes of thermokarst origin were taken into account. The boundary of the lake-lake area (the value of the lake-lake area is more than 1%) practically coincides with the 20-metre isohypsis at the section of the Viahtu Bay—Lakhtu Bay. Viahtu Bay—Lakh River. To the north it moves to a higher hypsometric level and from the latitude of M. Uanga to the latitude of

M. Lakh. Uanga to the latitude of m. Reidovyi runs along a 40-metre isohypsis. Further to the north the boundary again follows the 20-metre isohypsis. The territory with over 5% zonality is almost completely below 20 metre marks. The exception is the interfluve of the Pogibi—Vagis rivers, where it reaches 30 metres.

The main pattern of lake density ("zakazernosti") in the area, which is linked to the expanding river network on newly formed plains, is a uniform distribution of high-density zones along the coast at regular intervals of 10 and 20 km.

By analyzing changes in vegetation, climate, sediment composition, geological age, and the landscape of the coastal plain—while also considering the broader evolutionary trends of the Far Eastern coast—we can reconstruct the final stages of northwestern Sakhalin's geological history as follows.

In the Middle Wurm, a shallow water area existed in place of much of the coastal lowland. Diatom complexes from sediments exposed in the coastal escarpment between 4 m and 2 m absolute elevations correspond to lagoonal facies with varying degrees of desalination. The ancient shoreline probably occupied a position between the isolines of 1 and 5% of the shoreface. Diatom spectra dominated by Paralia sulcata from two 10-cm interbeds of grey clays overlying light grey fine-grained sands in a section of the 8–10 m long first supra-floodplain terrace of the Pogibi River (abs. 25–27 m), 16 km from the modern shoreline, indicate that these sediments were accumulated in the conditions of a desalinated lagoon. The reason for the rise of the Middle Jurm marine sediments to 25–27 m is the tectonic skew of the geomorphological surface in the direction of the coastline as a result of the growth of the Jolokor-Engizpal mountain uplift, the denudation slice of which for the neotectonic stage is 1.0–1.5 km.

The terminal phase of the Middle Würm was marked by climatic cooling and eustatic sea-level fall, resulting in the drainage of extensive coastal shallows and subsequent dissection of the emergent plains. The marine regression facilitated the progradation of substantial deltaic ridges at the mouths of large fluvial systems.

Sedimentation during the Late Würm epoch comprised alluvial, lacustrine-alluvial, and lake-alluvial deposits across the plain's surface, with a complex of fluvial terraces forming at higher topographic levels.

The transition to the warm, humid Holocene from the arid Late Glacial climate precipitated a key Early Holocene event: the initiation of peat accumulation on the northwestern Sakhalin coastal plain. This process was markedly enhanced during the thermal maximum of the Middle Holocene.

Climate warming in the Middle Holocene corresponds to the period of high sea level rise. According to A. M. Korotky and G. I. Khudyakov, on the western coast of Sakhalin Island the height of sea level rise relative to the present 5250 ± 90 BP was 3.0–3.5 m (Korotkiy 1990). On the coast of north-west Sakhalin Island, direct evidence of higher sea level rise during this period is found only near the mouth of the Viakhtu River. The age of the 3 metre peat bog at 5318 ± 320 BP (FESU-180) overlying estuarine-lagoonal sediments, taking into account the main patterns of coastal-marine sedimentation, allows it to be attributed to formations of the first rhythm of Holocene coastal-marine accumulation (Afanas'ev 1992).

Estuarine-lagoon sedimentation continued at lower hypsometric levels during peatland formation 5340 ± 540 BP (FESU-175); 5023 ± 445 (FESU-176).

The accumulation of a 0.5 m layer of aeolian sediments on the surface of the peat bog that began to form 7993 ± 807 BP (FESU-124) in the vicinity of M. Pogibi may be considered as evidence of high standing level Pogibi. Peat accumulation resumed after 5607 ± 700 BP (FESU-126) and continued until 4107 ± 432 BP (FESU-109), when the peat bog was again covered by aeolian deposits, but this time with a thickness of about 6 metres.

Subboreal climate warming initiated the phenomena of thermokarst and lake formation. Sea level rise during this period is probably associated with the beginning of the formation of thick aeolian sediment strata along most of the coast of the area.

At the mouth of the Pogibi River during this period there was accumulation of 2523 ± 360 (FESU-122); 3464 ± 403 (FESU-123) otorphic silts. Taking into account the peculiarities of the hydrodynamics of the estuarine zone of the tidal sea, the sea level position during this period was probably not higher than the present one, and if it was higher, it was insignificant, because during sisigian tides at surges the water level reaches the sampling points. In the north of the area during the second rhythm of coastal-marine accumulation, most of the Holocene accumulative protrusion—the Moskalvinskaya Pad tract and interval peat bogs on its surface—was formed. The beginning of the third rhythm of coastal-marine accumulation coincides with the beginning of the Middle Yaeya epoch. During this period, the position of the "Azov" type spit system stabilised, as evidenced by the dating of the first generation shaft of the Nyide spit −1470 ± 84 BP, the accumulative ledge in the area of Moskalvo, etc. was increasing. The subsequent slight decrease in level caused the cessation of estuarine-lagoonal sedimentation in most of the Viahtu River estuary at 940 ± 120 BP (FESU-179). Somewhat later, the lagoonal sediments were overlain by a low thickness (40 cm) peat bog 430 ± 82 BP (FESU-178). Synchronously with peat accumulation, a two-metre thick aeolian sand stratum 430 + ±78 BP was formed on the surface of the 20-m terrace (FESU-187).

Based on this coastal study, the following conclusions can be drawn. The Holocene evolution of the northwestern Sakhalin coast has been governed by its underlying morphostructure, whose principal elements have been inherited since at least the Neogene. Holocene coastal sedimentary formations are concentrated in areas with the greatest thickness of Neogene-Quaternary deposits.

Detailed analysis of the Upper Pleistocene alluvial-marine plain—whose appearance was significantly altered during the Holocene by peat accumulation, permafrost, and wind action—reveals that even a single geomorphological surface often has a complex, multi-stage origin involving multiple processes acting at different times.

In coastal areas with high sediment supply from land, particularly near estuaries, the development of wind-driven (aeolian) deposits is linked to periods of sea-level fall. Conversely, on coasts where sediment delivery is primarily controlled by marine processes, aeolian deposits form during phases of active wave-driven sediment accumulation. For example, the extensive dune systems of northwestern

Sakhalin began developing during the initial stage of the second major phase of wave accumulation.

2 Coasts with Holocene Accumulative Forms of Barrier and Free Type

The eastern part of the island existed as a eugeosyncline during the Palaeozoic and Mesozoic and is a fold-block zone of complex structure (Kharakhinov 1979a, b; Kharakhinov 2010). Tectonically, the coast of north-east Sakhalin is confined to a large trough that complicates the eastern side of the East Sakhalin anticlinorium (Fig. 5). The trough is a superimposed Cenozoic trough that was laid down in the early Early Miocene on a heterogeneous folded basement and has been steadily dipping throughout the Neogene, more intensely in the east and less significantly in the west. The trough is formed by a thick (up to 9000 m) thickness of Neogene terrigenous sediments assembled in gentle brachystructures.

The bedrock is represented by Upper Miocene siltstones, mudstones, sandstones, sands, clays and conglomerates of the Okobykai Formation (N13ok). These sediments comprise much of north-eastern Sakhalin and, in the area of coastal accretionary plains, are overlain by a cover of Quaternary sediments of varying thickness. The thickness of the Quaternary sediment cover ranges from 2–3 metres to 25–50 metres. Quaternary sediments are represented by a wide range of lithological varieties, with clastic soils predominating (in the upper part of the section). On the coast of the Sea of Okhotsk in the study area, Neogene bedrock sediments are overlain by Quaternary sediments of alluvial, alluvial-marine and marine origin, predominantly of Holocene age (m, am, vIV).

Among the lithological varieties of Holocene sediments composing the accumulative coast of the study area, coarse and medium sands prevail, occurring in the area of spits and overflows to depths of 10–15 metres. Below, as a rule, lagoonal-marine loams and clays of Holocene age are exposed, often with inclusion of organic remains. The coastal relief was formed under the combined action of abrasion and alluvial-marine accumulation processes.

During the neotectonic phase of Sakhalin Island's development, particularly in the north-eastern part of Sakhalin Island, large-scale sustained coastal and shelf deflections occurred, where brachyanticlinal folds were formed. Modern crustal subsidence rates here reach 2-5 mm/year.

In the Pleistocene time some depressions along the coast of the island experienced intensive deflections, which is recorded by the thicknesses of the corresponding sediments—up to 25–30 metres. The constant tendency for the coast and shelf of north-eastern Sakhalin to flexure is also evidenced by the character of coastal development. The latter is a classic type of lagoonal coastline, and it is possible to speak of inherited lagoonal development throughout the most recent phase of

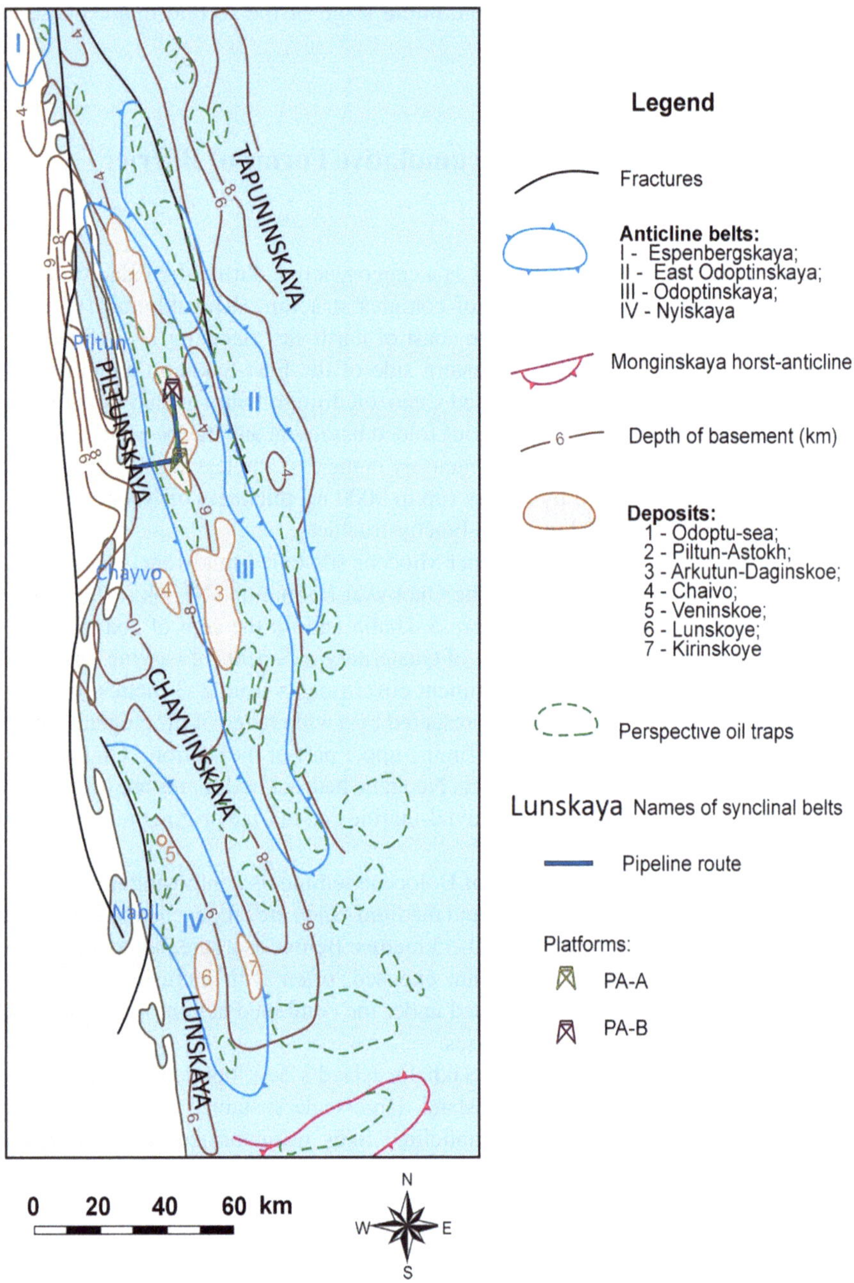

Fig. 5 Tectonic scheme of the north-east Sakhalin shelf (Kharakhinov 2010)

development. Modern crustal subsidence rates in this region are estimated to average 2–5 mm/year (Sergeev 1981).

The inherited development of the north-east coast of Sakhalin as a lagoonal coastline also suggests that the majority of the buried negative landforms identified during the geotechnical engineering surveys represent buried palaeolagoons filled with poorly compacted clayey sediments. These sediments are well identified on seismoacoustic profiles and have been confirmed by geological engineering drilling data from the area.

The area is one of the most morpholithodynamically active parts of the Sakhalin coastline. In addition to the constant southward current with a velocity of 10–20 cm/sec, much more active influence is exerted by reversible tidal currents with bottom velocities exceeding 90 cm/sec and by storm waves affecting the coastal slope to depths of 90–100 m. The relief of the submarine slope is characterised by a characteristic underwater slope relief. The topography of the submarine slope is characterised by sand ridges and hollows (Beloshapkov 2001). Sand ridges are dynamic landforms—their displacement rate can reach 25 m/year (Fig. 6).

Modern sediments are represented by fine-grained sand within the submarine slope from the cut-off to a depth of 14–15 m and clayey sand in the deeper part of the polygon. The thickness of modern sediments varies from 4 to 8 metres. They lie on the erosional surface of the Neogene Nutov Formation, represented by clays and loams with sand interlayers. At depths greater than 25 m, the sediments consist of well-sorted sands and siltstones with varying contents of shell fragments. At these depths, some areas of sediments with a gravel component were also noted.

Analysis of over two hundred histograms of the relative frequencies of granulometric distributions of bottom sediments has shown that the submarine coastal slope of north-east Sakhalin is characterised mainly by transitional conditions of bottom sedimentation (Fig. 7).

Lagoons are the most important element of the coastline of north-eastern Sakhalin. As a result of tides and the strong currents caused by them, the lagoons form drying facies (littorals) separated by erosional runoff troughs. Sediments fulfilling these troughs are mainly medium-grained sands (Rybakov 1996). Landscapes of all lagoons of the north-eastern coast have similar features, therefore the appearance of lagoon landscapes can be considered on the example of the previously studied Nabil lagoon (Brovko 1985). Within this lagoon several zones are distinguished by geomorphological features. These are drying areas within the inner lagoon littoral, which are the shallowest areas of the lagoon with average depths of 0.5–1.0 m, almost completely overgrown with sea grass Zostera marina and in much smaller quantities Zostera nana. Sandy-silt sediments predominate here. On the outer (relative to land) side of the drifts (supralittoral) a large amount of sea grass, which has been washed ashore and decomposes, accumulates. In Nabil Lagoon, at its northern tip, the supralittoral is up to 15 metres wide. Sediments of the upper littoral horizon are represented by black silty sands with a predominance of fine sand fraction. An anaerobic zone with hydrogen sulphide contamination is formed here. A similar zone is formed in the lower littoral horizon, the sediments of which are also represented by black silty sands with predominance of fine sand fraction. The middle

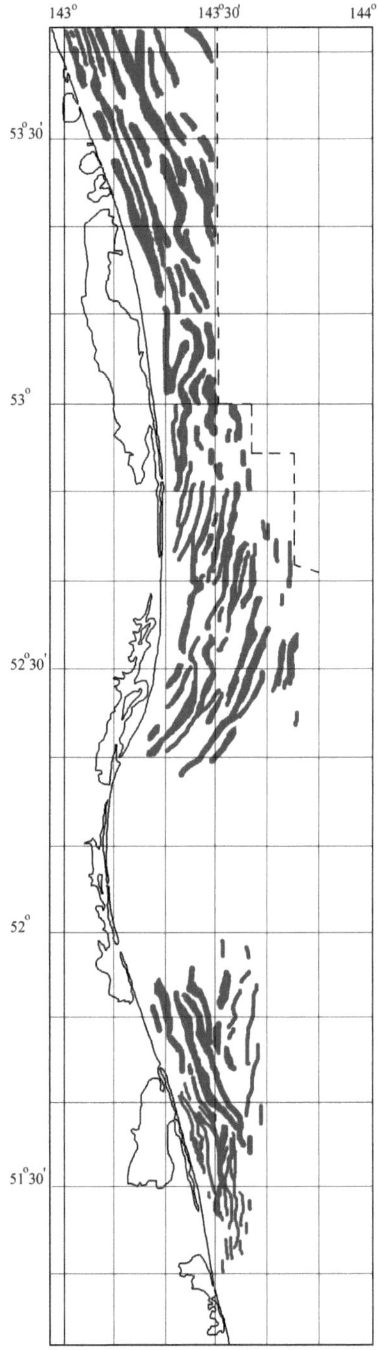

Fig. 6 Sand ridges on the submarine coastal slope of hollows (Beloshapkov 2001)

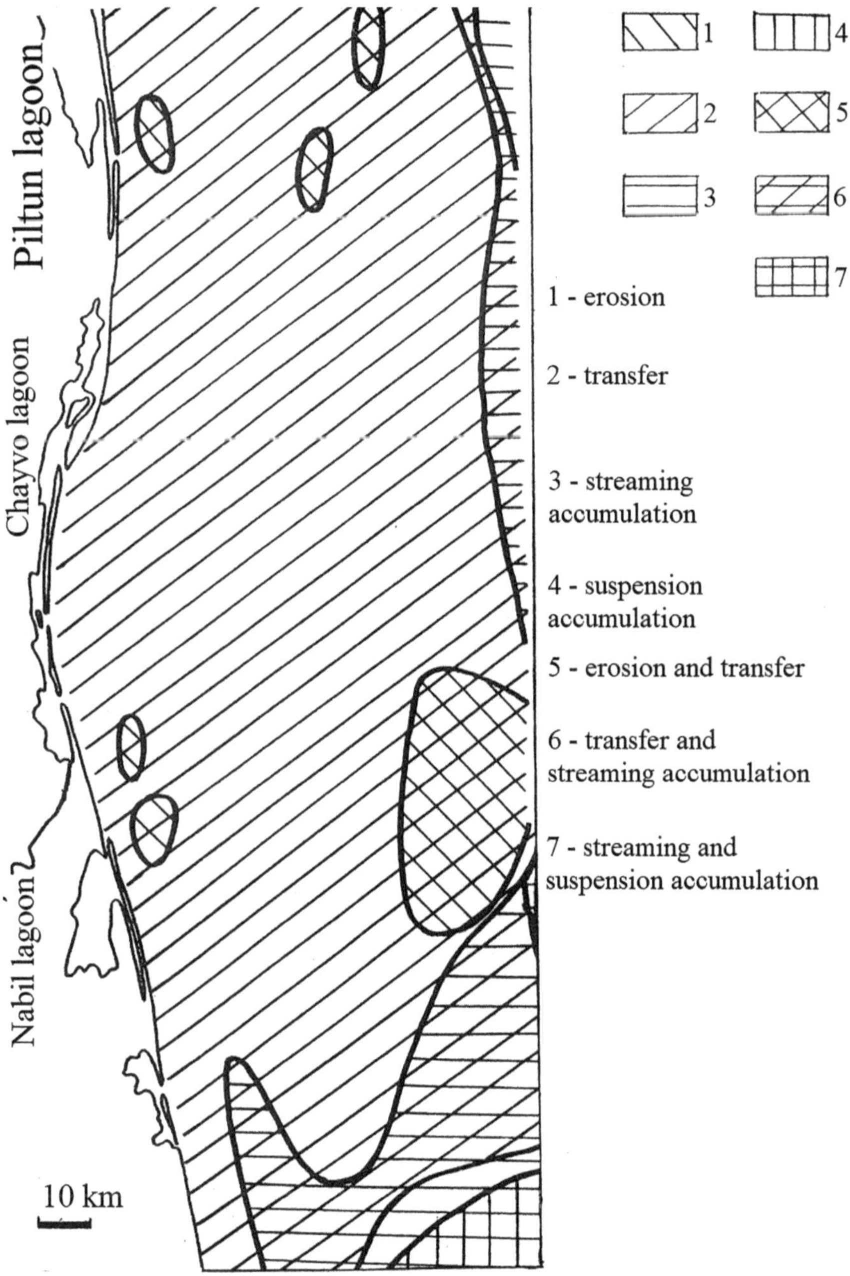

Fig. 7 Map-scheme of lithodynamic regimes (Rybakov 1989, 1991)

littoral horizon is composed of well-sorted medium-grained sands. Among the natural complexes of the closed lagoonal littoral, the lagoonal-mouth complex can be singled out as a special type. Such areas are located near the mouths of relatively large rivers flowing into the lagoons and are formed under the influence of their outflow of clastic material. The issue will be discussed in more detail in subsequent chapters.

According to the previously developed coastal development projection scheme, the coastal zone of north-east Sakhalin is stable with sea level rise and the amount of scour over the 50 forecast years is unlikely to exceed 50–100 metres, even with the observed sea level rise (Fig. 8).

2.1 Piltun Section

To the west of the Odoptinskaya anticlinal zone is the Piltun syncline, within the eastern wing of which the northernmost study area is located. This zone can be traced in the south-south-east direction from the area of the northern closure of Piltun Bay for a distance of more than 150 km. The width of the zone decreases to 5–10 km in the north and increases to 45–50 km in the south. The zone is complicated by longitudinal regional faults (Kharakhinov 2010). Within the coastline, a system of faults of Late Pliocene-Pleistocene age is distinguished.

The faults originated during the period of compressional tectonic activity that led to the formation of the Odoptinskaya anticline domes. The faults are represented by sublatitudinally oriented dips with dips both in the north and south directions and are confined mainly to the vault part of the southern dome in the depth interval of 500–2000 m. The faults merge at depth. The faults merge with depth and attenuate in the lower part of the productive section. In the upper 500 m of the section down to the seabed no tectonic disturbances are distinguished.

Geomorphologically, the study area is a complex coastal barrier separating Piltun Bay from the Sea of Okhotsk. Areas with high coastal scarps composed of Neogene-Pleistocene sediments (mainly unconsolidated sandy-pelitic fractions) alternate with Holocene overburden. There are no areas with high levels in the zone of Late Holocene migration of straits. They were destroyed here during the first phases of coastal-marine accumulation. Terrigenous rocks of Miocene-Pliocene age are exposed in the high coastal cliffs.

The deposits of the Nutov Formation (N1-2nt) are ubiquitously developed within the area under consideration and lie directly beneath the Quaternary formations. The contact with the underlying sediments of the Okobykai Formation is unconformable.

Two sub-horizons (sub-formations) are distinguished in the sediments of the Nutov Formation within the territory: Nizhnenutovsky (N1nt1). The horizon is represented by interlacing of sandy and siltstone reservoir beds containing oil and gas deposits and clayey sections-covers. The Verkhnenutovsky (N2nt2) horizon is

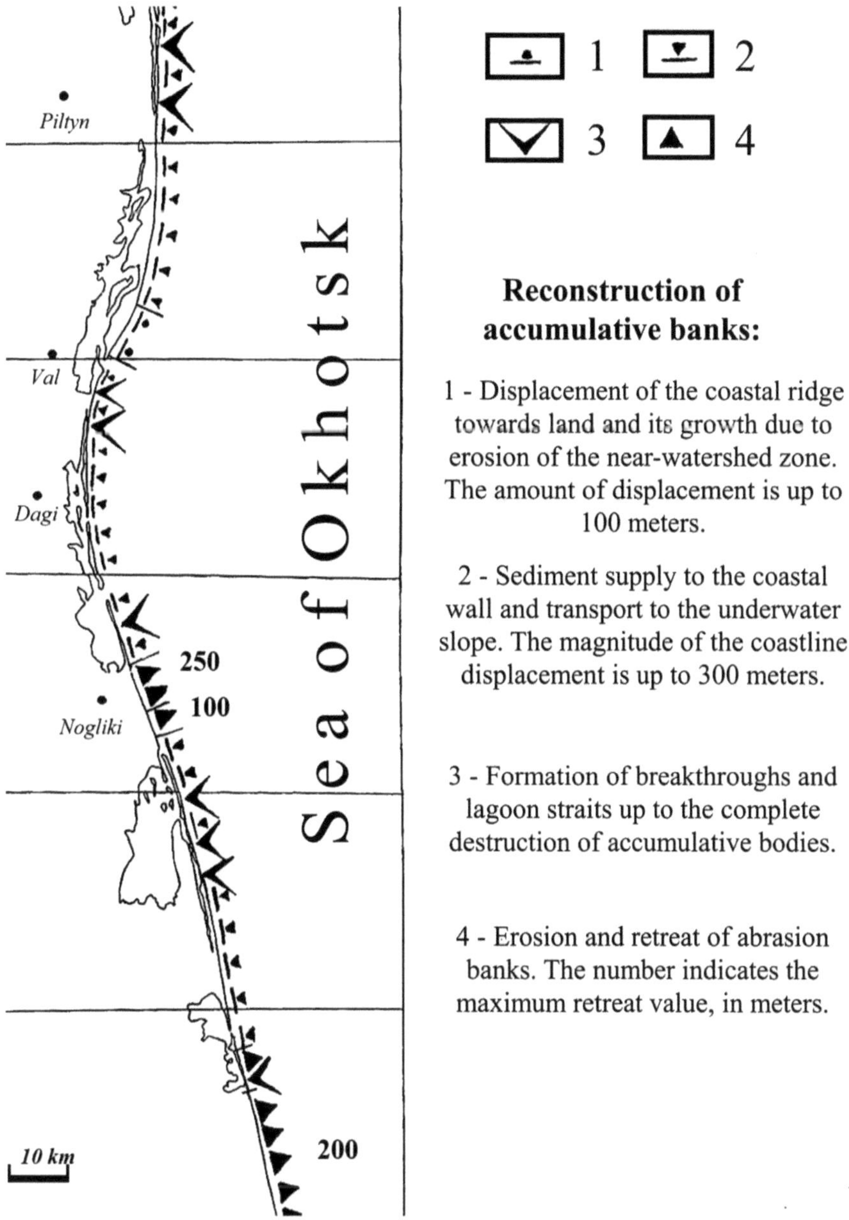

Fig. 8 Forecast scheme of coastal development under 1 metre sea level rise (Mikishin 1996)

composed of predominantly silty-clayey rocks with interbeds of siltstones and sandstones (Geology, USSR [Text]: M.: Nedra, 1970.—T. 33.—Part 1.—422 p.).

Sandstones are grey, clayey-grey, well-sorted predominantly fine-grained, often silty, less frequently medium- and coarse-grained, polymictic, clayey massive and platy thin-layered. The sands are identical to sandstones in appearance and mineral composition, differing only in the degree of lithification.

Siltstones and siltstones form separate low thickness (0.05–0.5 m, rarely up to 1–1.5 m) beds among sandstones and sands. They are grey, brownish-grey, less often brown, predominantly fine-grained, generally well sorted, less often with admixture of sandy-clayey material and charred plant detritus, platy, thinly layered.

Clays are composed of rare low thickness (1–2 m) beds of thin (0.05–0.1 m) interlayering with siltstones and sandstones. Gravelites and conglomerates lie mainly in the form of rare thin (0.1–1 m) lenses among fine-grained massive sandstones, often in association with nodules (0.1–0.7 m) of strong dense calcareous sandstones. The exposed thickness of Quaternary formations on Piltun Spit is 40 m.

Upper Quaternary marine sediments (mQIII) form a 7–12 m high marine terrace. The rocks are represented by interbedded (2–4 m) clays and multigrained (mainly medium- and fine-grained) sands containing low thickness (up to 0.5 m) lenses of gravel and pebble material. The total thickness of sediments is more than 20 m.

Modern marine sediments (mQIV) comprise coastal ramparts, interval depressions and beaches, as well as a low accumulative terrace. Beach areas, spits, embayments and bars are composed of well-sorted fine- to medium-grained sands and are mostly anchored by spiny sands. The thickness of Holocene lagoonal silts and peat bogs uncovered by boreholes exceeds 18 m (Fig. 9). At the same time, we obtained date (cal BP 9774-10,409 IGAS 5133) from the boreholes on the barrier form from depths of 29.9–30.0 and date (cal BP 12741–13,132 IGAS 5134) from plant remains from 30.0–31.0 m depth. However, already from the next depth interval of 31.0–35.6 m, dates (cal BP 34030-41,837 IGAS 5132) and (cal BP 40398-45,435 IGAS 5132) were obtained.

Modern aeolian deposits are observed everywhere. Dunes and avandunes are observed on high terraces in the form of low-powered hillocks, on Holocene areas of the overflow. The sediments are represented by fine and fine-grained sands, less often by dusty sands. The thickness of aeolian sands is small and averages 1.5–2.0 metres. Within the embankment the thickness of aeolian sediments is up to 4–8 metres.

According to the existing morpholithodynamic zoning scheme, the area starts from the strait connecting Urkt Bay to the sea and ends south of Astokh Bay (Fig. 10).

Fig. 9 Peat bogs exposed during erosion of the barrier form of Piltun Lagoon

Fig. 10 Scheme of morpholithodynamic zoning of north-eastern Sakhalin (Mikishin 2001)

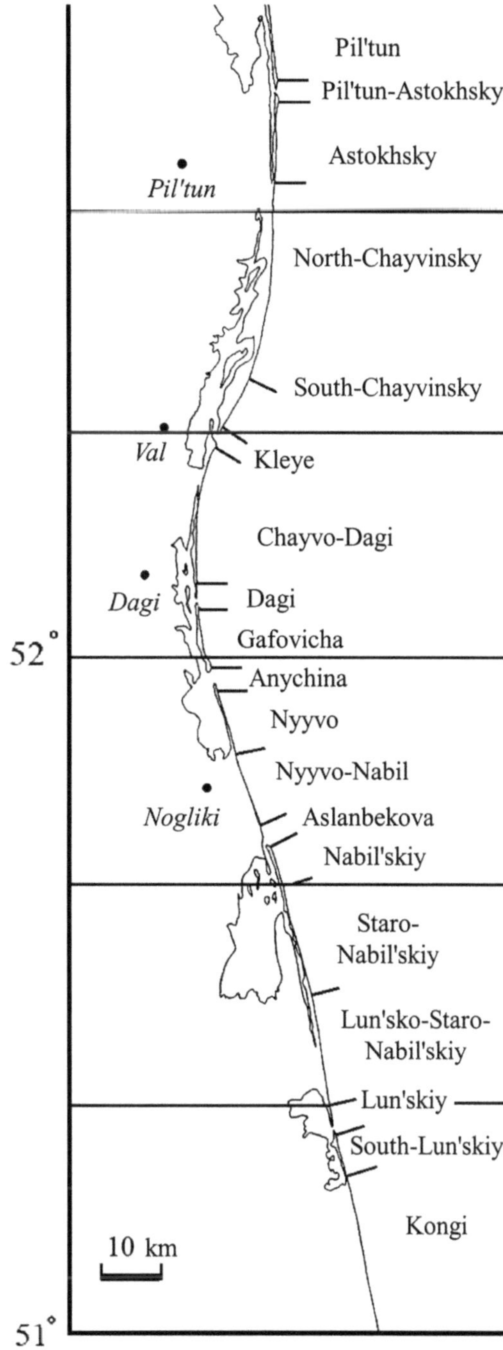

Fig. 11 Megafestons at the Holocene barrier form section

Trends in modern shoreline dynamics are determined by the megafaunal pattern of beach sediment distribution along the shore (Fig. 11).

The analysis of morpholithodynamics of complexly constructed and "classical" sandy embayments separating lagoons from the open sea has shown that complexly constructed barrier forms are characterised by a higher variability of deformations of the Holocene marine margin than fully Holocene embayments. Such barrier forms are also characterised by a greater variety of morphometric parameters of the upper part of the coastal profile.

Based on the results of beach sections and drilling of beach sediments, the thickness of the active layer, beach structure and bedrock bedding parameters (Holocene lagoonal sediments, Neogene clays and weakly consolidated siltstones) were determined. GPR data allow these parameters to be interpolated to the whole of north-eastern Sakhalin. The established thickness of the active layer, which reaches 4 metres, allows us to conclude that significant reorganisation of the coastal zone is possible. Seasonal restructuring of lithodynamic cells and trending changes in their planar position make the coastal zone vulnerable to erosion by high-frequency storm events along the entire coastline. Neogene terrigenous sediments composing high coastal scarps, including those of complex barrier forms, are represented by the Late Miocene Molasse Formation (Upper Miocene sediments of the Okobykai Formation (N13ok) [89]. Stratigraphically, it is a thick (up to 1500 m) regressive stratum of practically undiss tributed predominantly marine and coastal-marine, partly continental sediments. The lower part of the formation section is composed of irregularly interbedded clayey and sandy siltstones, argillites, weakly compacted clayey sandstones, and argillite-like clays. The upper part of the section exposed in the coastal escarpments is composed of sands with subordinate interbeds of clays, siltstones, weakly cemented sandstones, siltstones, with lenses of gravels, gravels, and in some places with rare interbeds of lignites (Fig. 12). Sands prevailing in the section are polymictic, horizontally and obliquely layered, different in granulometric composition, often with pebble and gravel inclusions.

Detailed geomorphological studies of the sediment convergence zone of north-eastern Sakhalin, located in the area of the Nyivo Lagoon barrier forms, have shown that it is the erosion of the coastal scarp in the 13 km section south of the Plastun

Fig. 12 Deposits of the Okobykai Formation (N13ok)) exposed in the coastal escarpment 300 m south of the Plastun Spit root zone

spit root (the southern element of the barrier form) that enables its development. The beach here has an incomplete profile and is composed predominantly of sands with a predominant fraction of 0.5–0.25 mm (47–73%). A significant content of gravel and pebbles (up to 55%) is determined by the erosion of Neogene gravel and pebble beds of poorly consolidated sediments in the coastal escarpment. The beaches of Plastun Spit have a full profile and the same granulometric composition. Avandune is observed in some areas. Dunes up to 15 metres high are observed at the southern tip of Gafovicha Island (northern element of the barrier form). In places, only isolated sand hills stretching along the shore remain, indirectly indicating a deficit of sediment in the coastal zone in these areas. At the base of the hills, the layering typical of coastal ramparts is clearly visible, with alternating layers of sand of varying coarseness, thin interlayers of gravel and, less frequently, small pebbles. In the upper part of the section of the hills, finer fractions prevail; gravel and pebbles are completely absent. The character of layering also changes sharply. Typical layering of marine coastal ramparts with distinct interlayers of sand and gravel at the top is replaced by oblique layering characteristic of aeolian formations. Quaternary sediments are thin (up to 20 m) and are represented in river valleys by alluvial formations, along the sea coast by marine and alluvial-marine sediments, and less frequently by aeolian sediments (in particular, on the barrier forms of Nõivo Lagoon).

Gravelly, coarse and medium coarse sands with insignificant content of dusty and clayey fractions are most widespread. Lenses and thin interlayers of loams are found in the sandy strata. Low Holocene marine terraces up to 2–3 m and 3–5 m high on the shores of the Nyivo Lagoon are composed of peat bogs of different composition (Fig. 13).

Fig. 13 Holocene peat bogs in the scour ledge of the second marine terrace of the lagoonal shoreline

Fig. 14 Section of the first marine terrace of the lagoon bank

The deposits of the modern (first marine terrace) and laida on the shores of the Nyivo Lagoon are represented by peaty-gley deposits with sand interlayers (Fig. 14).

Relative elevations of the underwater relief of the inner delta formed by tidal currents and waves penetrating into the water area of the lagoons reach 1.5 metres. Morphometric parameters of the inner lagoon beaches are presented in Fig. 15. The position of the rear edge of the inner lagoon beaches is on average 2 m lower than on the beaches of the offshore part of the spit (Fig. 16).

Fig. 15 Morphometric parameters and beach structure at the erosion site of the second marine terrace (Baltic height system)

2.2 Nabil-Lunsky Section

The overflow of the Nabil Lagoon and the terrace that continues southwards between the Nabil and Lunjska lagoons is a genetically unified form. Closely related to it are the overflow of the Lunska Lagoon and the accumulative terrace extending southwards from it. The shoreline is straightened. Barrier forms can break through, in which case new straits are formed. Figure 17 shows a strait formed in 2011 that has existed for 7 years.

The surface of the predominant part of the accumulative forms is fixed by moss-grass cover and shrub vegetation. Woody vegetation appears on the older parts of the accumulative forms, along with which large areas of scattered sands with extensive deflation pits and remnants of scattered dunes are often found in the areas of dune massif development. The ends of the embankments near the lagoon channels are also devoid of vegetation.

The site of detailed morpholithodynamic observations is a Holocene accumulative marine terrace adjacent to an ancient coastal escarpment developed in the Upper Miocene sediments (Okobykai Formation), represented by

Fig. 16 Morphometric parameters and beach structure on a section of high marine terrace with an overgrown coastal escarpment (Baltic height system)

Fig. 17 Newly formed strait at Old Nabil Embayment

pebble-gravel-sandy-clay formations and on the surface of the abrasion-denudation level of the Okobykai Formation sediments. The same sediments are currently exposed in an active coastal scarp at the root of Plastun Spit.

The width of the mid-late Holocene coastal marine plain in the detailed study area is 1015 metres (excluding the beach zone), of which 375 metres is a residual lagoonal overgrown water body and 640 metres is a zone of coastal ramparts with an avandune zone traced along the entire shoreline, the height of which exceeds the height of adjacent coastal ramparts and aeolian surf terraces by 1–4 metres (Fig. 18).

The width of the Middle Holocene terrace decreases uniformly from north to south. If in the axial zone of the site it is, as noted above, 1015 m, then 1 km to the north it is already 1105 m, and 1 km to the south it is 860 m. The thickness of gravel and sand deposits in the zone of the coastal ramparts, according to drilling data, is about 11–15 metres. The aeolian breakwater deposits are underlain by heavy grey loams of lagoonal facies with hydrogen sulphide odour. The absolute mark of the

Fig. 18 Coastal-marine Late Holocene terrace (1973 Aerial photography)

Fig. 19 System of actively eroding dunes

lagoonal deposits roof in the avandune zone is −11 metres (BS 77). A characteristic feature of the terrace at this site is a system of watercourses crossing it in the NW-SE direction, through which the runoff from the backwater lakes that emerged during the formation of the Middle Holocene marine terrace and, in the north of the site, from the residual lagoonal overgrown water body is carried out. The distance between the watercourses is approximately the same and is 900–1200 metres.

The terrace is adjacent to an ancient coastal escarpment developed in Upper Miocene sediments (Okobykai Formation) represented by sandy-clay formations. The height of the coastal embankments reaches 6–8 metres. A system of dunes up to 20 metres high is observed at the northern boundary of the site, which are actively eroded at present (Fig. 19).

Deflationary processes, according to the 1952 ASF analysis, were active even before the period of intensive geophysical works accompanied by the laying of roads and glades.

A low modern terrace up to 4 m high adjoins the avandune from the seaward side. The avandune shows traces of modern erosion (Fig. 20).

In all morphological elements of the beach the medium sand fraction predominates, from the back edge to the edge of the beach its fraction decreases from 45 to 35%. The content of fine sand, the second largest fraction, varies along the profile from 26 to 33% with a maximum content in the middle part. Coarse sand reaches a maximum content of up to 30% in the cut-off area. Gravel and pebble size of beach sediments is less than 1%, reaching its maximum value in the area of the cut-off.

As noted by predecessors, the coastal scarp has been stable over the last few decades (Buzlaev 1982; Mikishin 1996). The coastal zone rather had accumulative tendencies in this area, as evidenced by the present-day, low eolian-surf terraces attached to the avandune, on which in turn avandunes are beginning to form (Fig. 21).

The formation of a new megafeston system and destruction of terraces and avandunes was probably triggered by the reorganisation of lithodynamic cells as a result of cofferdam construction several years ago (Figs. 22, 23, 24 and 25). The

Fig. 20 Avandune with scour marks

Fig. 21 Modern marine terrace up to 50 metres wide, formed after 1952

cofferdam protective structure has now been dismantled. When comparing the 2015 and 2012 COPs, we were able to identify zones of coastal scarp failure that became active after the construction and operation of the cofferdam (2012–2013). At the same time, the maximum rates of destruction during this period were noted in the areas of stable coastline for the last 60 years. However, it is unknown how long and with what consequences the coastal system will come to a state of dynamic equilibrium.

Interestingly, sands of the size observed on the beach were actively mobilised by winds of more than 15 m/s directly from the splash zone (Fig. 26).

Within 30 min, a tree about 30 cm thick was covered with wet sand at the foot of the beach ledge.

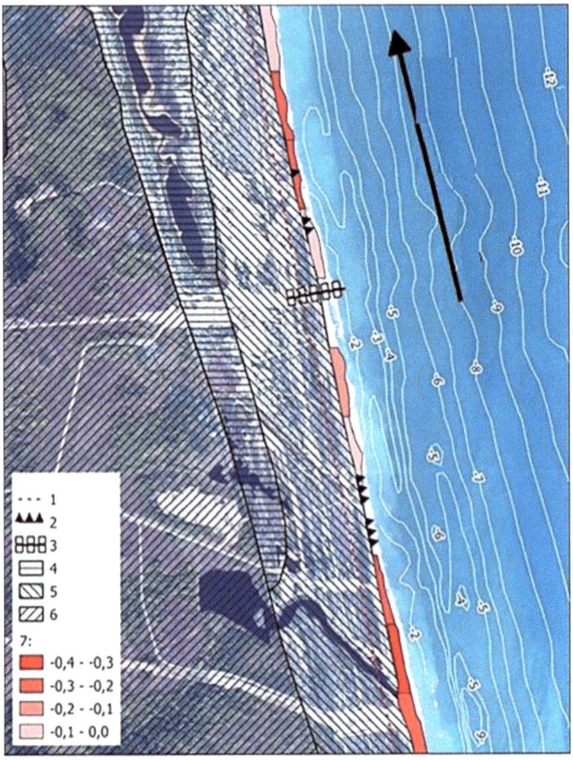

Fig. 22 Scheme of morphology and dynamics of the coastal zone. Notation: 1—surface composed of Neogene rocks with submerged lakes; 2—Holocene lagoonal and residual-lagoonal surfaces; 3—dune massifs on the surface of middle-late Holocene accumulative aeolian-surge formations; 4—middle-late Holocene coastal ramparts; 5—coastal scarp erosion (values are signed on the scheme); 6—avandune and modern terrace

Fig. 23 Coastal zone structure at the beach promontory site

Fig. 24 Scour of modern terrace in the pipeline area

Fig. 25 Avandune scour on the southern flank of the site

3 Abrasion-Denuding Shores of Subarctic and Temperate-Cold Seas

3.1 Middle and Southern Part of Eastern Sakhalin

The geological structure of the territory includes rocks of a wide age range from the Upper Cretaceous to the Holocene. In the northern section, the shores are composed of terrigenous, siliceous-terrigenous tuff-siliceous and, to a lesser extent, volcanogenic rocks of Cretaceous-Palaeogene age and are resistant to denudation processes. The Berezovskaya Formation (K2br) sediments are the most widely represented in the area (Geology, USSR [Text]: M.: Nedra, 1970.—T. 33.—Part 1.—422 p.). Relationships with the underlying formations of the Bogatinskaya Formation are tectonic throughout.

The suite combines a thick thickness of terrigenous formations (siltstones and sandstones) with olystostrom horizons, including lenses and blocks of tectonised volcanogenic-siliceous and intrusive rocks.

Fig. 26 Wind transport of foam and wet sand in the beach area

Fig. 27 Coastal scarp and cecures in bedrock of the Berezovskaya Formation

Cliffs are active, frost weathering and scree processes are evident. The beaches are narrow boulder-pebble and boulder clasts, abrasion cliffs are widespread (Fig. 27). The width of the marine basement terrace (benches) reaches 2 km. Thus, the coastline under consideration can be attributed to the abrasion-denudation type.

Cenozoic formations are represented mainly in the south of the territory and compose the stratigraphic section of the water area bottom. Lateral differences in the composition and structure of the Cenozoic strata are mainly facies in nature due to the sequence and stage of formation of the sedimentary basin of the Sea of Okhotsk.

The Lyukaminskaya Formation (P2lk) forms the seabed in the middle part of the area to depths of 5–10 metres. It is overlain by a thin cover of modern marine sediments. It is also exposed for a small length in the coastal escarpment south of the

Fig. 28 Coastal scarp with mudstones of the Pileng Formation at its base

Vengeri River. The Formation is represented mainly by sandstones and siltstones with interbeds of mudstones, coal mudstones, coals and, to a lesser extent, conglomerates and gravelites.

The Pilenga Formation (P3pl) overlies the sediments of the Lyukamin Formation (Polunin 1992). It is represented predominantly by clayey-siliceous siltstones and mudstones with interbeds of opaque siltstones and mudstones, less frequently diatomites, gravelites, and conglomerates. The Formation deposits are exposed in the coastal escarpment (Fig. 28).

They are overlain by undivided Middle and Late Pleistocene sediments several metres thick and are presumably of alluvial-marine origin. Middle-Late Pleistocene sediments fulfil estuarine extensions of rivers and are represented by bars and fragments of coastal ramparts. In the southern part, the banks are composed mainly of weakly consolidated rocks. Scour intensity is up to 1.5 m per year. The shores composed of rocks of this type are precipitous. The shallow part of the shelf is an abrasion-accumulative plain with separate ridges.

The underwater slope is gentle with underwater ramparts. The leading relief-forming factor on the coast is landslide, rockfall, erosion and abrasion processes. Subsidence landslides reach several hundred metres (Fig. 29). Mudflow cones are observed in the mouths of small rivers. The magnitude of bank scarp retreat is 1.5–2.0 metres per year. According to our data, the intensity of bank erosion has increased approximately 1.6 times over the last 20 years in the area under consideration.

Basement terraces are also widely represented on the opposite, western shore of southern Sakhalin Island. Discussion of the age and geospatial parameters of Pleistocene terraces on the coast of Sakhalin Island has subsided somewhat recently

Fig. 29 Block landslide

(Svitoch 2004). Nevertheless, the divergence of views on the problem of marine cyclic geomorphological levels has not been overcome here and is due, on the one hand, to the lack of results of age dating of sediments and, on the other hand, as noted by A. N. Alexandrova, to the fact that no micropalaeontological evidence in favour of a marine origin has been found for terraces 50–80 m and higher (Alexandrova 1976).

However, the actual geomorphological "trace" in the formation of these surfaces remains practically unattended by researchers. Our studies of the formation and development of rocky shores and, in particular, benches had a great influence on the formulation and presentation of such an approach to analysing the results of high sea level rise in the Late Pleistocene (Afanas'ev 2020, Afanas'ev and Uba 2022). Here we also draw attention to the fact that basement terraces on the sea coast with neotectonic movements of positive sign are nothing but palaeobench overlain by marine, estuarine-lagoonal, alluvial, slope sediments of different types and eolian sediments. The thickness of these sediments on one cyclic abrasion-denudation level, depending on the position in relation to the source and type of sedimentation, and the peculiarities of coastal scarp destruction, can vary in a very wide range. For example, even along a 3 km section of the bank scarp in the vicinity of Novoselovo settlement, the thickness of the proluvial-colluvial cover of the basement terrace changes from 3 to 20 metres as the exposed section of the bank scarp moves away from the source of slope deposits.

We analysed the composition of the sediments overlying the paleobench, made high-precision measurements of the geospatial position of the geological section elements and the bedrock basement, and for the first time obtained age determinations from the coastal marine sediments of the so-called Boshnyakovskaya terrace, the age of which was previously determined as Middle Pleistocene (Aleksandrova

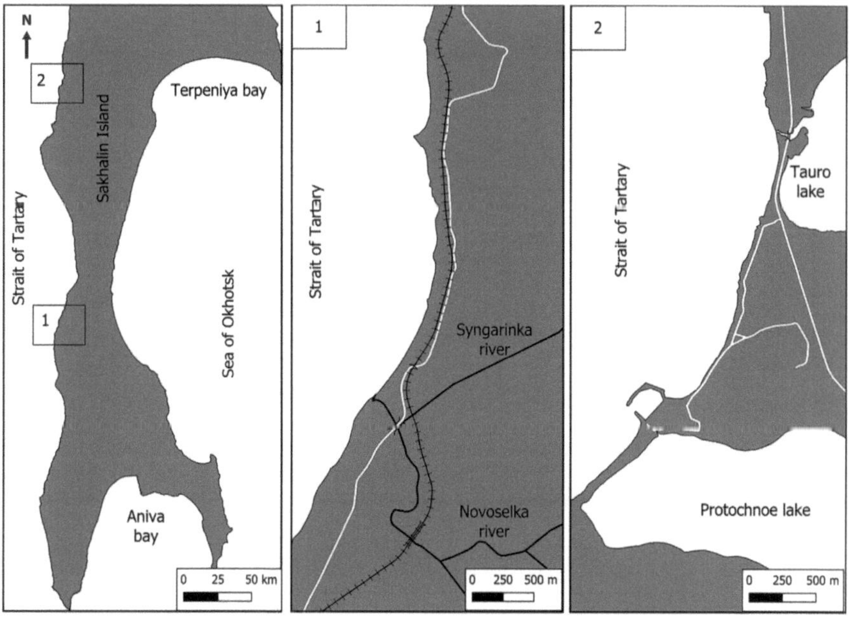

Fig. 30 Location of the study area. 1—marine basement terrace in the vicinity of Novoselovo settlement, 2—marine basement terrace in the vicinity of Shakhtersk settlement

1982; Kulakov 1973; Korotkiy et al. 1997). The location of the study area is shown in Fig. 30.

All measurements were made by SOUTH Galaxy G1 GNSS receivers (GPS, GLONASS, BEIDOU, GALILEO). Pickets at the working sites were taken in RTK (real-time kinematics) mode from the local base station (BS) and using Topcon GTS236-N total station. Coordinates of local BSs were obtained in static mode from 2 permanent BSs of EFT-CORS network located in Korsakov and Yuzhno-Sakhalinsk. RTK accuracy is within 1 cm in plan and 2 cm in height, static accuracy is not more than 2 cm in plan and 2 cm in height. ITRF2014 (epoch 2010.0) coordinates of BS Korsakov and Yuzhno-Sakhalinsk, height above WGS84 ellipsoid were taken as reference ones. The obtained pickets heights were converted to heights relative to the geoid EGM2008 (Earth Gravitational Model 2008).

The study of the geological structure of the terrace in the vicinity of Novoselovo settlement showed that the soil layer is underlain by hard gravelly sandy loam with rare thin clay interlayers (Fig. 31). The thickness of the upper layer varies within 4.0–4.5 metres. There are no water intrusions in this layer in the outcrop of the of the bank scarp.

The groundwater level in the wells is below the bottom of this horizon, which is defined by GPR sounding as the first complex. The sediments are characterised as slope deluvium overlying alluvial marine sediments of Upper Pleistocene age. Below this layer, a 2.5–3.5 metre layer of waterlogged gravels can be traced in the outcrop of the coastal scarp. In the boreholes located closer to the ancient bank

Fig. 31 Orthophotomap of the coastal escarpment in the vicinity of Novoselovo settlement: 1–3 geological sections with sampling for radiocarbon and facies analyses, 4–5 points for determining geospatial parameters of the paleobench

Fig. 32 Aquifer over estuarine-lagoon facies

scarp, this layer has a lower thickness and a higher hypsometric position than in the bank cliff. Probably, here we are dealing with a pack of marine beach sediments proper, and taking into account the proximity of the river these sediments are most likely alluvium, reformed by wave surfing processes near the estuary zone. It is this layer in its lower part that is the zone of linear water seepage in the coastal scarp (Fig. 32).

Table 1 Results of absolute age determination by radiocarbon method

Lab. number	Location, sample type	Altitude position, m	Radiocarbon age, years	Calibrated age (calendar)
LU-10099	Section №1, wood	35.00–35.19	25,880 ± 340	30,200 ± 390
LU-10100	Section №1, algal peat	34.86–34.90	25,120 ± 290	29,440 ± 310
LU-10715	Section №3, wood	35.09–35.00	24,410 + 430	28,630 ± 480
LU-10718	Section №3, wood	35.60–35.66	25,420 ± 280	29,650 ± 290
LU-10717	Section №3, wood	35.18–35.24	25,590 ± 350	29,830 ± 390
LU-10716	Section №2, wood	32.56–32.70	25,790 ± 370	30,080 ± 430

Further below, a subhorizontal horizon of coastal-marine sediments, estuarine-lagoonal lithofacial complexes—clays and loams, less frequently loamy clay loams of bluish and grey colour, in which mixed complexes of brackish-water and marine diatom algae were determined and an interlayer of intensely tiniferous gravel and pebble sediments with a thickness of 2.04 m lying between them. Absolute age determinations were obtained from these sediments (Table 1).

The roof of this layer in the description section № 3 of the coastal scarp lies at 35.09 m, in №1 at 35.19 m, and in wells located 50–70 metres from the exposure of this contact in the coastal scarp towards the ancient coastal scarp the height marks of this layer are 35.89 m, 36.2 m. The thickness of the upper layer is 0.8–0.91 m. The bottom of the lower layer of silty sediments with thickness of 1.38 m (№2) lies at the mark of 31.76 m.

Below the clayey layer there are gravels and gravels with sandy and sandy loam and sandy loam and very dense gravelly sandy loam 5.50 m thick (Fig. 33). They lie directly on the marine abrasion-denudation basement, which is, in fact, a geomorphological cyclic surface, benched in the rocks of the Kholmskaya (N_1 hl), Nevelskaya (N_1 nv), Chekhovskaya (N_1 ch), Verkhnedui (N_1 vd), and Kurasiyskaya (N_1 kr) formations of Upper Pleistocene age. Absolute surface elevations of the palaeobench are 25.26 m in area 5 and 25.64 m in area 4 (BS 77).

The abrasion-denudation cyclic geomorphological level in the Shakhtersk area is developed in siltstones, mudstones and fine-grained sandstones of the Kurasi Formation (N1kr) of varying strength (Fig. 34). Absolute elevations of the basement surface lie in the range of 6.0–6.5 m (BS 77). The palaeobench is directly overlain by sandy-pebble sediments, the layered nature of which allows them to be attributed to marine beach sediments. These sediments are overlain by aeolian fine-medium-grained sands 2.0–3.0 m thick. A fragment of palaeodune about 5 m high is observed on the terrace surface.

Interestingly, the roof of the estuarine-lagoonal sediments filling the palaeoview to the south of the considered section of the basement terrace lies at approximately the same elevations as the palaeobench surface. The thickness of aeolian deposits overlying the surface of lagoonal silts is 7–8 metres.

As a result of the study of the basement marine terraces of the western coast of Sakhalin, the precise geospatial parameters of the basement and overlying

Fig. 33 Tintinned gravel and pebble deposits beneath lagoonal sediments

sedimentary strata have been determined for the first time. The age of marine regressive marine terrace sediments, defined by the predecessors as early Pleistocene, was significantly rejuvenated as a result of radiocarbon dating of these sediments. Thus, the amplitude of the neotectonic movements, for the estimation of which the analysis of marine terrace levels is used, is significantly lower than the inferred ones (Bulgakov and Senachin 2019). The facies analysis of the sediments and the geospatial position and surface parameters of the terrace basement allow us to categorise these formations as palaeobenchs. A very interesting fact is the inclination of the surface of benches to the general flow lines, which were bays and straits during the period of sea level lowering. This phenomenon is also found on pumice abrasion shores of Holocene age (Afanas'ev 1989).

The morpholithodynamics of the coastal zone of the coastal area under consideration is poorly studied (Goloudin 1980). Systematic observations of the

Fig. 34 Orthophotomap of the coastal scarp in the vicinity of Shakhtersk settlement

Fig. 35 Wave protection retaining wall

development of coastal processes in the study area have not been carried out for over 40 years, despite repeated attempts to consolidate the coastal scarp (Fig. 35).

In relation to the problem of longshore sediment migration, the morpholithodynamic situation can be characterised as follows. On the coastline as a whole, there is fairly intensive coastal erosion and sediment migration against a background of active wave action. In the benched areas, longshore migration in the beach zone occurs mainly as a result of the superposition of surges of various genesis and high tides. The main volume of sediment in the benches moves in the depth range 4–9 metres (calculated closure depth).

The coastline of Terpeniya Bay from Tikhiy Bay to Senyavina Bay is located in the south-western part of Terpeniya Bay in the Sea of Okhotsk (Figs. 35 and 36).

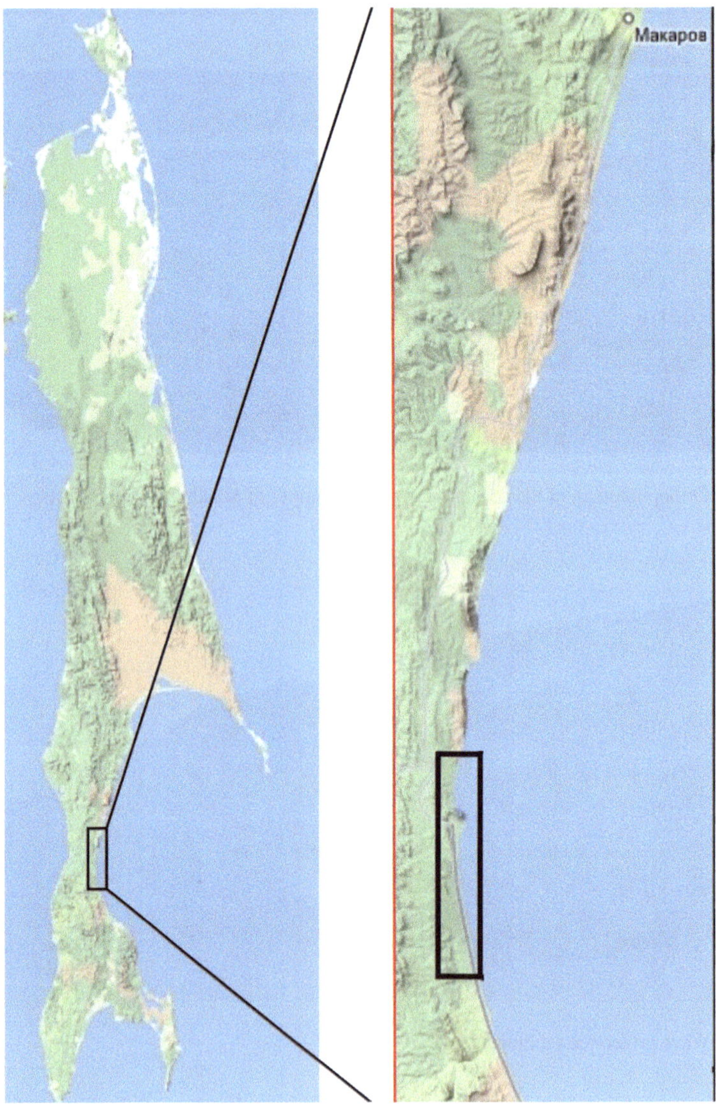

Fig. 36 Coastline of Terpeniya Bay from M. Bay Tikhiy to Bay Senyavina

The shoreline forms an arc with the ratio of the depth of incision to the distance between the bounding headlands of 0.13. The distance along the traverse of the capes is 52 km, depths 30-40 metres. The 10-metre isobath is 0.1–1.5 km from the coastline, the 50-metre isobath is 7.0–10.0 km away.

From the west, the coast is bounded by the South Kamyshov Ridge, whose watershed runs 7–17 km from the coastline. The eastern seaward spurs and slopes of the ridge come close to the coast and are cut through by a number of short rivers and watercourses. The total number of rivers, including unnamed watercourses of 3–4-km length from Vzmorie village to Starodubsky village, is more than 25. Their estuaries cut through the coastal zone every 1–2 km. The largest rivers include such rivers as Naiba (about 50 km), Ai, Firsovka, Chernaya, Dudinka, Aidar and Baklanovka.

A narrow strip along the shore, and in some places fragmentarily to the slopes of lowlands, is joined by a Holocene marine terrace up to 3–4 m high and a 5–8-metre high Late Pleistocene terrace.

River mouths are blocked by longshore storm ramparts, except for the mouth of the Naiba River. A wide delta is developed here, complicated by the Lebyazhie and Torfyanoye lakes and along the shore narrow lagoons - Utinoe, Dolgoe, Maloe Pridorozhnoye.

Relief-forming rocks of the coast are represented by loose sediments of Quaternary age, which are developed everywhere. Abrasion-denudation coastal slopes of lowlands are composed of sandstones and siltstones of Nutovka and other formations of Neogene age. The promontories are composed of intrusive rocks or siliceous shales, tuffs and quartzites, which are relatively resistant to abrasion.

From the mouth of the Manui River to the Mulovskoye River, small in length (200–700 m) abrasion-denudation sections of the coast alternate with accumulative sections. The abrasion-denudation shores have cliffs 20–50 m high, complicated by slope processes, and narrow sandy-pebble beaches. Accumulative shores are confined to the valleys of small rivers, where low (2–5 m) terraces composed of grey viscous clays are widespread. On the Mulovsky Cape, which protrudes more than 1 km from the general shore line, the high (more than 100 m) and steep shore escarpments of the Cape, developed in Miocene effusives, predominantly light yellow tuffs, are now almost unabraded and represent typical denudation slopes. Woody-rubble sediments, massively formed by weathering, descend along them to narrow boulder-pebble beaches. A very wide benches in rocky rocks are developed here. North of the cape, its top reaches 450 metres, and south-east, where abrasion has cut narrow ridges (apparently steeply falling rock strata) –700 metres.

Between Vzmorie and Starodubskoye, the coastline presents a system of marine accumulative terraces, among which the most clearly distinguished are: modern, including the beach and an adjacent series of coastal ramparts 0–2 m high; Holocene –3–4 m high; and Late Pleistocene –10–15 m high. The first two are composed predominantly of sands, less frequently of gravels and pebbles; the third is composed of dense blue or bluish-grey clays overlain by thin layers of brown clays or peat bogs (Mikishin 1991).

In the Vzmorie-Beregovoye section, the width of coastal accumulative terraces varies mainly from 0.5 to 1.0 km, widening up to 1.5 km only in the river valleys. Accumulative processes prevail: sandy beaches 30–50 m wide are developed almost everywhere [96]. Scouring is noted only to the south of Vzmorie settlement and to the north of the mouth of the Aidar River, where the accumulative terrace is destroyed. Active scarps 2 m high are developed in clays and peat bogs. A breakwater wall made of logs was built near Vzmorie settlement to protect the road along the terrace. Steep sloping beaches 30–40 m wide, composed of well-washed brownish-yellow coloured gravels or sandy-pebble material, depending on the season, are common in the eroded areas. Interseasonal deformation of beach sediments at the rear edge is 4–6 metres.

Between the settlements of Beregovoye and Kirpichnoye, a shore with a dead cliff and a leaning accumulative terrace is observed in a section of about 10 km. The cliffs are developed in very dense clayey sands of the Maruyama Formation. The attached marine terrace is only 100–300 m wide. The shoreline here is sediment deficient, as evidenced by scour scarps 1.4–2.5 high in the low accumulative terrace, stretching almost continuously for 9 km. Erosion of the terrace occurs rarely, mainly during strong autumn storms, since the shores are protected by wide (50–70 m) sandy beaches of a full profile with rare coastal ramparts.

South of the settlement of Kirpichny the coastal accumulative plain widens considerably (up to 5–10 km). Most of it is swampy. In some places there are chains of low dunes developed on ancient coastal ramparts. The shore belongs to the accumulative lagoonal type.

The sediment load measured during the study of the Sakhalin GRES intake bucket is 40–50,000 m^3/year (repeated measurements on hydrodynamic cycles).

Analysis of the calculated wind energy characteristics, beach parameters and the structure of the underwater relief makes it possible to characterize the morpholithodynamic situation in the area between Cape Senyavina and Cape Tikhii as follows.

Throughout the entire length of the coast under consideration, the component of the wave energy flux directed along the normal to the shore is 15–20 times higher than the values of the longshore component. This value can be further increased due to quicksand waves. The calculated parameters of the longshore wave energy flux capacity are rather high and amount to about 10 thousand tonnes/season per 1 m of the coastal profile. The resulting longshore wave energy flux is significantly lower and only in a few sections reaches 5–6 thousand tonnes/season per 1 m of the profile (Afanas'ev 2015).

Considering the spatial structure of the longshore wave energy flux with mean annual parameters, two sections can be distinguished. The first one extends from Senyavina Bay to the mouth of the Ai River. The resulting wave energy flux here is directed to the north. At the coastal section from the mouth of the Ai River to the Tikhogo Bay, the resultant wave energy flux is directed southwards. In the Dudino area, the average annual sediment load in the surf zone is about 0.5 million tonnes per season. In order to appreciate the scale of magnitude, the following comparisons can be made. At present, the beach sediment reserves of the coastal zone from Vzmorie to Starodubskoe in summer are about six million tonnes, while during the operation of the Dudino quarry about four million tonnes have been removed from the coastal zone.

When analysing the seasonal variability of the characteristics of the longshore component of the wave energy flux, two modes were established. In summer months the wave energy flux is directed northwards practically along the entire coastline. Its energy supply in this direction decreases. The mesorelief features of the submarine coastal slope, established as a result of analyses of the APS from different years, confirm the northern direction of the sediment flow in summer. In autumn the flow is southwards. Energy availability decreases from north to south. In general, for the section of the coastline of Terpeniya Bay under consideration. From May to July, a weakening of the wave energy flux is characteristic, and from August to November - strengthening.

When analysing the changes in the calculated wind energy parameters of the coastal zone over a multi-year time interval, it was established that in the period 1970–1984, some weakening was observed, and from 1985 onwards, wave activity has been increasing, which is consistent with the regional trends we have established (Afanas'ev 1994).

Well-sorted coarse-grained, dynamically mature sands predominate in the study area, showing a "normal" granulometric cross-sectional structure with a predominant concentration of larger fractions in the upper part of the beach. The distribution of values of statistical characteristics along the shore showing the longshore structure of sediment composition indicates a depressed character of accumulation on the coastal section under study, with a minor role of sediment load from rivers and streams flowing into the sea.

References

Afanas'ev V.V. To a question about the morphostructure of north-western Sakhalin / V.V. Afanas'ev // Geological and geomorphological features of some regions of the Far East and the zone of transition to the Pacific Ocean. - Vladivostok: DVPI, 1989. - P. 100-105.

Kharakhinov, V.V. New data on the geological structure of the Sakhalin and Lower Priamurya junction zone / V. V. Kharakhinov, I. M. Alperinovich, V. M. Nikiforov, A. A. Tereshchenkov // Reports of the Russian Academy of Sciences. - 1979a. - T. 245. 1. - P. 192-195.

Kononov, V. E. Prospect of oil and gas bearing capacity of north-western Sakhalin in the light of new geological and geophysical data / V. E. Kononov, R. M. Makarov, V. M. Nikiforov et al. // Pacific Geology. - 1986. - № 3. - P. 57-61.

Mishakov, G. S. Peculiarities of the North-Sakhalin sedimentary basin development / G. S. Mishakov, V. S. Kovalchuk, M. S. Yaroshevich // Sedimentary basins and their oil and gas content. - M: Nauka, 1983. - P. 81-84.

Khudyakov, G. I. Geomorphotectonics of the South of the Far East (Theory Issues) [Text] / G. I. Khudyakov - M.: Nauka, 1977. 256 p.

Svitoch, A. A. Marine Pleistocene of the Russian coasts (palaeogeography) / A. A. Svitoch // Oceanology. - 2003. - T. 43. - №. 5. - P. 783-794.

Solovyev, V. V. Relief development and formation of Quaternary sediments of Sakhalin / V. V. Solovyev, G. S. Ganeshin // Proc. of VSEI, Nov. ser. All-Russian Geological Institute, Nov. ser. - T. 172. - M: 1971. - 152 p.

Alexandrova, A. N. Stratigraphy of Quaternary sediments of Sakhalin Island / A. N. Alexandrova // Materials on stratigraphy and palaeogeography of East Asia and the Pacific Ocean. - Vladivostok: Far Eastern Scientific Centre of the Academy of Sciences of the USSR, 1987. - P. 50-69.

Afanas'ev, V. V. Evolution of the Far Eastern Seas Coast in the Holocene / V. V. Afanas'ev // Coastal Evolution in the Conditions of Ocean Level Rise. - M.: Izd-vo MSU, 1992. - P. 166-174.

Aleksandrov, S. M. Sakhalin Island [Text] / S. M. Aleksandrov. - Moscow: Nauka, 1973. - 182 p.

Aleksandrova, A. N. Pleistocene of Sakhalin [Text] / A. N. Aleksandrova. - Moscow: Nauka, 1982. - 292 p.

Yanagida, M. The process of formation of Holocene river terraces on the Saru River / M. Yanagida // Tirigaku heroi: Vol. 54. 1981. - Hokkaido Island: Per. from Jap. - I-01967. - P. 423-435.
Korotkiy, A. M. Exogenous geomorphological systems of sea coasts [Text] / A. M. Korotkiy, G. I. Khudyakov. - Moscow: Nauka. 1990. - 216 p.
Kharakhinov, V. V. Oil and gas geology of the Sakhalin region [Text] / V. V. Kharakhinov. - Moscow: Nauch. Mir, 2010. - 275 p.
Kharakhinov, V. V. New data on the geological structure of the Sakhalin and Lower Priamurya junction zone / V. V. Kharakhinov, I. M. Alperinovich, V. M. Nikiforov, A. A. Tereshchenkov // Reports of the Russian Academy of Sciences. - 1979b. - T. 245. 1. - P. 192-195.
Geology, USSR [Text] : M.: Nedra, 1970. - T. 33. - Ч. 1. – 422 p.
Sergeev, K. F. Modern vertical movements of the earth crust of Sakhalin Island / K. F. Sergeev, Y. P. Nikitenkov, V. A. Shulman // Dokl. of the USSR Academy of Sciences. - 1981. T. 257. - №. 1. - P. 202-205.
Beloshapkov, A. V. Accumulative forms of the coastal zone and shelf of northeastern Sakhalin / A. V. Beloshapkov, A. I. Gordin, V. V. Ilyin, V. F. Putov // Mankind and the coastal zone of the World Ocean in the XXI century. - Moscow: Geos, 2001. - P. 126-140.
Rybakov, V. F. Lithodynamics and sedimentation in the Chaivo lagoon / V. F. Rybakov // Geographical studies of sea coasts. - Vladivostok: DVGU. 1996. - P. 105-110.
Rybakov, V.F. Lithodynamics of the Okhotomorskiy shelf of Sakhalin Island / V.F. Rybakov // Coastal zone of the Far Eastern seas. - Leningrad: Publishing House of the USSR State Committee, 1991. - P. 85-97.
Brovko, P. F. Lithological and geomorphological conditions of development and types of littoral of Nabil lagoon (Eastern Sakhalin) / P. F. Brovko, P. F., A. I. Kalin. F., A. I. Kafanov // Benthos and conditions of its existence on the shelf zones of Sakhalin. - Vladivostok: Far Eastern Scientific Centre of the Academy of Sciences of the USSR. 1985. - P. 20-25.
Mikishin, Yu. A. Forecast of the development of the lagoon coast of Sakhalin Island at the ocean level rise / Yu. A. Mikishin // Geographical studies of the sea coasts. - Vladivostok: DVGU, 1996. - P. 66-75.
Mikishin, Yu.A. Geomorphology and lithodynamics of the shelf and coastal adjoining areas of the pipeline (Sakhalin-1 project) / Yu.A. Mikishin, P.F. Brovko, // FEFU report. -Vladivostok, 2001. – 107 p. - (Funds of "Sakhalin Research and Development Institute of Marine Oil")
Buzlaev, V. A. Report on aerogeological and engineering-geological studies of coastal cliffs to study the intensity of abrasion processes and the forecast of landslide phenomena within the north-eastern coast of Sakhalin for 1978-1981 / V. A. Buzlaev. - Yuzhno-Sakhalinsk, 1982. - 79 p. - (Funds of "Sakhalin Research Institute of Marine Oil")
Polunin, G. V. Map of lithological complexes and exogenous processes of the land and shelf of Sakhalin on the scale of 1:500000 / G. V. Polunin. - Khabarovsk, Yuzhno-Sakhalinsk, 1992.
Svitoch A. A. Marine Pleistocene of the Far Eastern coasts of Russia // Pacific Geology. - 2004. - T. 23, № 3. - P. 76-93.
Alexandrova A. N. Marine terraces of Sakhalin and Hokkaido // Shelf Geomorphology and Palaeogeography. - 1976. - M.: Nauka. - P. 123-131.
Afanas'ev V. V., Uba A. V. Destruction of Rocky Coastes of Cold Seas // Processes in GeoMedia. - Springer, 2022. - Vol. V. - P. 243-252. (Book series: Springer Geology). – DOI:10.1007/978-3-030-85851-3_27.
Bulgakov R. F., Senachin V. N. Marine terraces and the influence of the hydroisostasy effect on vertical movements of Sakhalin // Geosystems of transition zones. - 2019. - T. 3, № 3. - P. 277-285. https://doi.org/10.30730/2541-8912.2019.3.3.277-285.
Kulakov A. P. Quaternary coastlines of the Sea of Okhotsk and the Sea of Japan. - Novosibirsk: Nauka, 1973. - 189 p.
Korotkiy A. M., Pushkar V. S., Grebennikova T. A. Marine terraces and Quaternary history of the Sakhalin shelf. - Vladivostok: Dalnauka, 1997. - 229 p.
Vetrovoy isthmus of Iturup island – holocene strait / V. V. Afanas'ev, E. I. Ignatov, A. V. Uba, N. N. Dunaev, I. O. Leontiev, A. O. Gorbunov // IOP Conference Series: Earth and Environmental

Science. – IOP Publishing, 2019. – Vol. 324, №. 1. – Article Number 012029. DOI:10.1088/1755-1315/324/1/012029

Goloudin, R. I. Substantial composition and dynamics of coastal-marine sediments of the western coast of Southern Sakhalin / R. I. Goloudin // Oceanology. - 1980. - VOL. XX. - Issue. 3. - P. 507-512.

Mikishin, Yu. A. Geomorphology of the shores of Terpeniya Bay / Yu. A. Mikishin // Coastal zone of the Far Eastern seas. - L.: Izd-vo GO USSR, 1991. - P. 98-104.

Afanas'ev, V. V. Geomorphological aspects of the problem of Sakhalin Island coastal protection / V. V. Afanas'ev // Geomorphology. - 2015. - № 2(182). - P. 28-37.

Afanas'ev, V. V. Coastal erosion of southeastern Sakhalin / V. V. Afanas'ev, V. A. Buzlaev, M. A. Rudavets // "Northern Pacifica". - Vladivostok: 1994. - P. 15-20.

Patterns of Coastal Destruction at the Present Stage

V. Afanas'ev Viktor

Abstract Based on the processing of aerial photography (flights in 1952, 1974) and space photography (2014 and 2015) in a GIS environment, followed by the analysis of the spectral density graphs of erosion intensity, the parameters of the erosion of the Sakhalin coast were determined and the spatial structure of it was characterized. In 2011, in the area of the town of Nevelsk, Sakhalin Oblast, we began instrumental observations along a network of 100 benchmarks concreted into the surface of an argillite-siltstone bench coseismically uplifted as a result of the earthquake on August 2, 2007, in the area of the town of Nevelsk, Sakhalin region.

As shown by a comparative analysis of the rates of destruction of benches and coastal ledges in rocks of weak and medium stability, the destruction of the latter occurs at rates approximately an order of magnitude greater than the rates of destruction of the surface of the coseismically uplifted bench. The maximum contribution to the destruction of the bench is made by frost weathering at negative air temperatures during the open sea period. The number of freezing and thawing cycles during this period in the conditions of the tidal sea reaches 100–150 cycles.

Keywords Sakhalin Island · Spatial structure of the erosion · Erosion rhythms · Morphometry of the beach · Megacusps. Coseismic coastal uplift · Shore platform · Sea cliff · Erosion rates

1 Introduction

Based on GIS processing of aerial photo information (1952 and 1974 flights) and space photo information (2014 and 2015) with subsequent analysis of spectral density plots of scour intensity, the parameters and spatial structure of erosion of the coasts of northeastern Sakhalin were determined and characterised. The main

V. Afanas'ev Viktor (✉)
Sakhalin State University, Yuzhno-Sakhalinsk, Russia

T. Chaplina (ed.), *Coastal Geosystems of Cold Seas Under Conditions of Climatic and Anthropogenic Changes*, Earth and Environmental Sciences Library, https://doi.org/10.1007/978-3-032-12247-6_3

regularities of the spatial structure of the destruction of these shores determine the phenomena of rhythmicity in the manifestation of scour processes. Scour rhythms with wavelengths of 1.3; 2.0; 4.0; 5.6; 6.5; 11.6; 25.6; 64 km have been established.

Mean annual scour rates do not exceed 2 m/year anywhere. Maximum scour rates were observed in areas that represent high (14–28 m) Pleistocene terrace levels. These sites are also characterised by the maximum volumes of sediment entering the coastal zone each year when a 1 m long bank scarp is eroded (5.6–7.7 m^3).

The high rhythmicity of scouring processes (in our case ≈ 30–60 km) on the levelled shores of north-eastern Sakhalin is associated with stable zones of divergence of sediment fluxes, characterised by a reduced volume of beach sediment. The 1.3 km scour rhythmicity corresponds to the dimensionality of large megaflood beach structures.

Long-term studies of beach structure and morphometry for practical design purposes on the shores of Sakhalin Island have led to the conclusion that a fully wave-absorbing profile is formed on sand and gravel beaches with a sediment volume of 70–90 m^3 per 1 metre of beach length. On most of the accumulative shores of Sakhalin Island there is no sediment deficit, however, scour can be observed even at average beach sediment volumes within lithodynamic cells of 140–150 m^3/m.

As demonstrated in three sites with significant sediment supply on the upper elements of the coastal profile, scouring of these shores is due to the uneven distribution of beach material along the shore as a result of the formation of megafloods and their systems.

Rocky coastal erosion is a function of bedrock characteristics, wave climate, tidal regime and weathering patterns. The relative contributions of marine and subaerial processes, as well as geological properties of eroding rocks, have been and are discussed in many dozens of studies. Detailed analyses of the status of this area of coastal geomorphology are presented in analyses (Stephenson 2013; Sunamura 2015; Trenhaile 2016a, c).

Wave action is currently considered to be crucial in the development of such coasts in areas of high wave energy, while the role of weathering is more significant in warm temperate and tropical regions Cruslock 2010). Nevertheless, the nature and relative effectiveness of physico-chemical and biological processes on benches, the influence of climate, wave regime, tidal range, rock characteristics and the spatial and temporal development of coastal platform destruction cannot be unequivocally characterised to date (Trenhaile 2019).

2 Destruction of Banks Composed of Bedrock

The development of geospatial technologies has now made available high-precision photogrammetric digital multi-temporal elevation models of abrasion-denudation shores of cold seas, which make it possible to determine the rates and nature of fracture of bedrock of weak and medium stability. New high-precision data on fracture rates of coseismically uplifted benches as a result of the Nevelsk earthquake on

2 August 2007, $M_w = 5.2$ in the town of Nevelsk allowed us to establish that the destruction of subvertical shore ledges in rocks of weak and medium stability occurs at rates about an order of magnitude greater than the rates of destruction of subhorizontal benches. The role of wave action in the development of benches is mainly confined to the mobilisation and transport of debris derived from frost weathering at sub-zero air temperatures during the open sea period. The number of freezing and thawing cycles during this period reaches several dozens. Highly accurate ortho photos and 3D models of the high coastal escarpment composed of volcanogenic-sedimentary rocks made it possible to identify the failure mechanisms of cliffs with heterogeneous strength and determine their quantitative parameters (Fig. 1).

Abrasion is the major modern morpholithodynamic process on rocky shores, comprising approximately 75% of the coastline (Emery 1982). Accurate and continuous monitoring of bedrock bank erosion is virtually impossible, so averages over periods of in-situ observations of cartographic and remote sensing surveys are used to determine abrasion rates (Sunamura 1992).

However, even long-term average rates of abrasion of bedrock banks should be used with caution in hazard determination and coastal zoning. It is well known that the January 1983 storms in California washed away about 14 m of a cliff of Miocene siltstone that had retreated at an average rate of 0.2 m/year between 1931 and 1982 (Griggs 1985).

We observed a similar phenomenon more recently in the vicinity of the Prigorodnoye LNG plant, located on a basement terrace composed of weakly cemented mudstones and siltstones of the Bykov Formation (K2bk) of the third stability class (Polunin 1984). A single storm in autumn 2017, 4–7 meters of the coastal ledge were destroyed, while over a more than ten-year period the coast was eroding at a rate of 0.1–0.15 m/year (Fig. 2).

Urgent observations of rocky shoreline failure were also carried out between 1999 and 2007 by the Centre for State Monitoring of the Geological Environment at SakhGRE on the coast of Terpeniya Bay and the Tatar Strait.

In Terpeniya Bay, the retreat of the low basement terrace margin composed of Neogene siltstones of the Kurasi Formation N1Cr was documented (Fig. 3).

The average rate of scarp retreat over the observation period was 0.26 m/year per linear metre of shoreline (Figure 4a). However, it should be noted that, as with the Prigorodnoye LNG landfall, the main erosion event occurred as a single event in October 2005. The contribution of this storm event to the total amount of abrasion over the entire observation period was 77%.

On the Sakhalin shore of the Tatar Strait in the vicinity of Ilyinsky, measurements of cliff scarp retreat were made at three measurement sites (Figure 4b–d).

Despite the fact that the coastal scarp here is composed of lithified sandstone-siliceous rocks of the Kholmsky lithotype of the fourth stability class, the average annual rates of cliff face retreat are approximately the same as the rates of cliff failure developed in rocks of the II and III lithotypes (Polunin 1992). The average annual retreat rates of the cliff face are 0.25 m/year for Site № 1 and 0.32 m/year for Site № 2 (Fig. 5). The exception is Site 3, where the average annual rate of cliff retreat was 0.05 m/year per 1 linear metre of bank scarp.

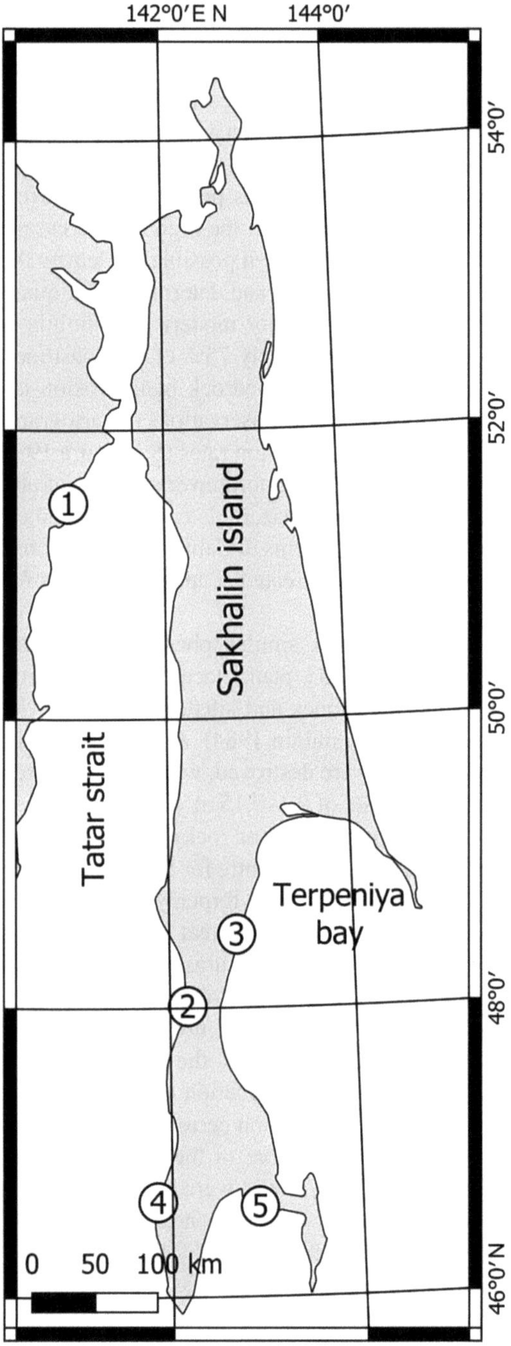

Fig. 1 Map of actual rocky shore study material: 1–high coastal scarp of De-Kastri Bay; 2—cliff of Tatar Strait; 3—scarp of the basement terrace of Terpeniya Bay; 4—benches near Nevelsk; 5—coastal scarp near Prigorodnoye LNG

Fig. 2 Coastal escarpment in the Prigorodnoye LNG area

Fig. 3 Active cliff excavated in N1 (Cr) deposits

The area with the lowest rates of shore scarp destruction in the Tatar Strait is located in the wave shadow of the small harbour structure for the most energetically provided wave of northern rhumb waves (Fig. 6). In our opinion, the maximum contribution to benches failure is made by frost weathering at negative air temperatures during the open sea period. The number of freezing and thawing cycles during this period reaches 100–150 cycles.

Regarding the issue of stability of coasts composed of bedrock, the following fact should be noted. In 2012, a slope-type bank protection was constructed for the

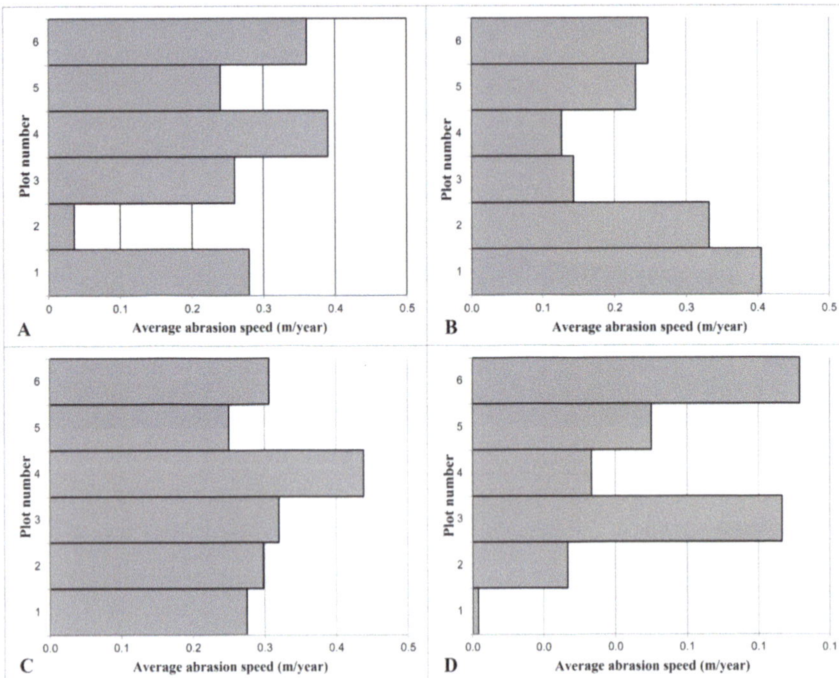

Fig. 4 Average abrasion speed: (**a**) basement terrace of Terpeniya Bay (0.26 m/year); (**b**) Tatar Strait Average, site № 1 (0.25 m/year); (**c**) Tatar Strait, site № 2 (0.32 m/year); (**d**) Tatar Strait, site № 3 (0.05 m/year)

Fig. 5 Coastal scarp (site №2) open to swell of prevailing directions

motorway section near Vzmorie settlement. Natural stone of dacite composition of the Middle Miocene age was used as the main wave-absorbing element. The use of natural stone in marine hydraulic engineering construction implies its ability to resist multiple alternating freezing and thawing in a water-saturated state without visible signs of destruction. According to DCS 5-84 "Departmental Construction

Fig. 6 Coastal scarp (site No. 3) in the wave shadow of the harbour structure

Standards", the frost resistance of such stone should be not less than FR100. However, volcanic rocks (dacites) turned out to be very unstable to weathering processes: already in 2013 the monoliths cracked and the bank protection was destroyed (Fig. 7).

Differentiation of wave impacts and subaerial weathering factors can be achieved by studying the destruction of coastal platforms—benches formed on rocky shores with poor rock stability. An exceptional opportunity in this respect was given by studies of the destruction of the benches coseismically uplifted by 0.8–1.0 m as a result of the Nevelsk earthquake on 2 August 2007, Mw = 5.2 in Nevelsk (Fig. 8). The amplitude of bentsch uplift in the area of Nevelsk was 90–100 cm (Fig. 9).

The main structural-formational zones are shown in Fig. 10. Immediately at the observation site, the bench is composed of greenish-grey fine-grained, clayey, weakly cemented sandstones, replaced along the strike by siltstones and thin interbeds of stronger calcitised fine-grained sandstones.

Conditional designations: 1—Upper Miocene Kurasi Formation: lower sub-formation (kr1)—siliceous mudstones, opoks), and upper sub-formation (kr2)—sandstones with siltstone interbeds; 2—Middle Miocene Upper Duy Formation—interbedded sandstones, siltstones and mudstones, coal beds; 3—Lower Miocene Nevelskaya Formation: lower sub-formation, (nv1)—sandstones and upper sub-formation, (nv2)—sandstones with interlayers of siltstones; 4—Oligocene-Lower Miocene Kholmskaya Formation, (hl)—siltstones with interlayers of sandstones; 5—Eocene-Oligocene Arakai Formation, (ar)—sandstones with interlayers of basalt tuffs; 6—axes of synclines (a) and anticlines (b); 7—faults (a), including those with presumed discharge (b) and vasthrust components of displacements (c); 8—directions of horizontal displacements; 9—orientation of layering (a)

Fig. 7 Destruction of natural stone of dacite composition of bank protection structure due to low frost resistance

Fig. 8 Site of benches raised by the earthquake (7 days after the event)

and fault displacements (b); 10—direction of compression; 11—points of structural observations.

In 2011, in the Nevelsk area of the Sakhalin Oblast, we started instrumental observations using a network of 100 reporters concreted into the benthic surface. The distance between the profiles is 25 metres, the distance between the repers along the profile is 10 metres. The first repeater is located on the moribund edge of the Bench. In 2019, measurements were carried out only on 32 repers (Figs. 11 and 12).

Some of the benchmarks were lost as a result of ice impact in winter 2019, some benchmarks were filled with anthropogenic soil or destroyed.

As an example of changes, Fig. 13 shows a visual series of the Bench surface dynamics at one of the sites (5/2) of transect № 5 / rep 2.

Since July 2016, measurements have been made by subtracting multi-temporal height field maps of 2.25 m^2 plots with the repeater in the centre. The height field

Fig. 9 Fencing blocks of spoil areas at Bench during maximum high tide; (**a**) 12.07.2007, (**b**) 14.06.2007, after the earthquake

maps were constructed by digital photogrammetric method in Agisoft Photoscan software. An aluminium frame with dimensions of 1.5 × 1.5 metres, set horizontally using an optical leveller, is used for model referencing. At the four corners of the frame there are markers with pre-measured coordinates encoded as a circular binary code. The frame is orientated using a bussol to ensure correct positioning of the axes in space. The SONY A6000 camera (24 megapixels) is used for taking pictures. Next, Agisoft Photoscan software performs the initial linking of photos with the construction of a cloud of common points, points with low reliability of coordinates determination are removed from the cloud. Then on the basis of the remaining reliable points the camera calibration is specified and the dense cloud and polygonal model are built (Fig. 14).

The next step is to build a height field model with a planning resolution of 0.001 m. A matrix of height differences over the period allows the calculation of the specific volume of damage per 1 cm^2 of area and, accordingly, the average velocity over the site. Figures 15, 16 and 17 show the areas subjected to intensive direct sea impact.

Figure 18 describes an area 50 metres from the edge of the raised benches that is eroded primarily by weathering, provided its products are removed by the superposition of surge and tide. Areas outside the zone from which the clastic material is washed away erode at minimal rates.

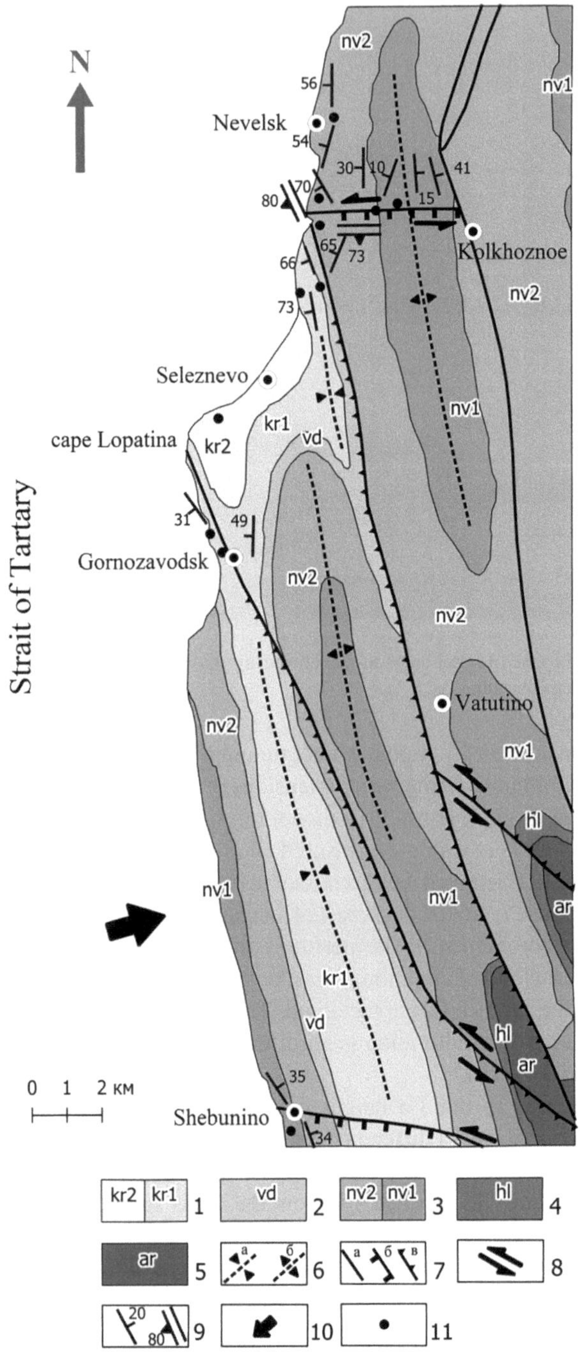

Fig. 10 Geological and structural scheme of the Nevelsk area (Golozubov 2012)

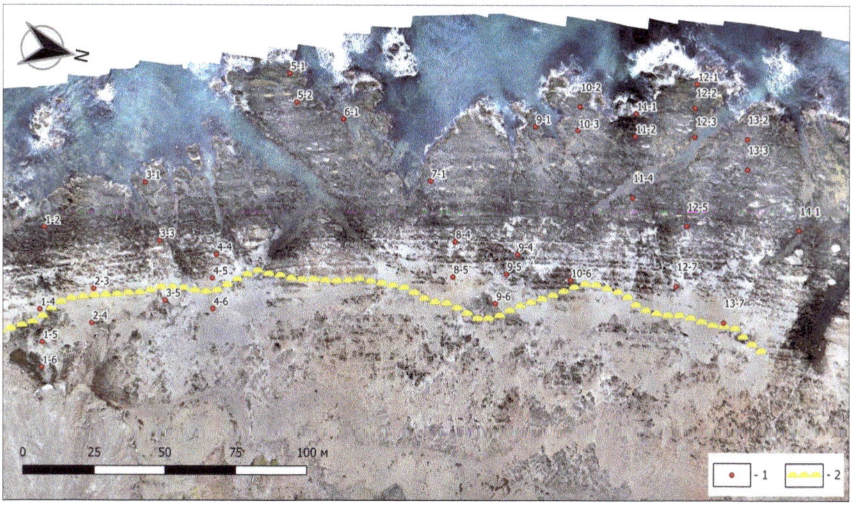

Fig. 11 UAV image from 2018: 1—position of the landmarks used until 2019; 2—the rampart formed at the edge of the ice in 2019

Fig. 12 Shaft formed at the ice edge 2019

Fig. 13 Visual series of Bench surface dynamics at one of the sites (5/2) borehole № 5/reach 2

The minimum benches surface fracture rates of up to 0.25 cm/year are observed in areas overlain by clastic material washed off the moribund part of the benches during the superposition of maximum sea level and wave levels (Fig. 19).

The maximum benches destruction rates were observed at the north-western boundary of the raised benches, in the vicinity of stations 11–13, where already in

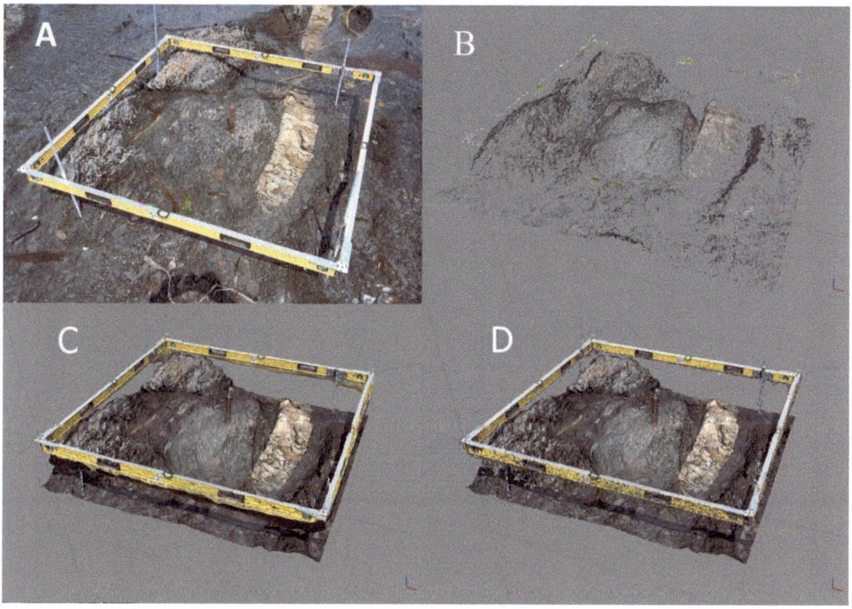

Fig. 14 Sequence of polygonal model building: (**a**) photo of the reference with the set frame, (**b**) connecting points, (**c**) polygonal relief model, (**d**) dense cloud of points

2018 some of the 35 cm deep repers were washed out, i.e. the average annual benches destruction rate exceeded 5 cm/year. In areas periodically flooded by water, the rate of benches surface destruction reaches 4.37 cm/year (Fig. 20).

In addition, it was found that the siltstone fracture rate was about 2 times higher than the fracture rate of calcitised nodules (Fig. 21).

Calcitised nodules with smooth, undisturbed surfaces at 40 m from the edge of the benches are eroding at rates of 1–2 mm/yr., which corresponds roughly to the middle of the interval of measured values worldwide (Trenhaile 2016b; Trenhaile 2016a, c). Figure 22 shows that the destruction of the nodule surface is related to the structural features of the rock.

Laboratory experiments using rocks from eastern Canada showed that erosion rates for different rock types were maximum at the low sisigian tide level (which experiences two wetting cycles every 24 h) (Fig. 23). Above and below this maximum, velocities decreased due to more infrequent changes in conditions (Porter 2010; Trenhaile 2016a, c). Very important for our study is the recent finding in a study of Canadian coasts that maximum fracture rates at high hypsometric levels are likely related to frost weathering, ice exposure, and other mechanisms (Trenhaile 2016a, c).

High fracture rates of layered, highly fractured Bench rocks, which are about an order of magnitude higher than those measured in temperate climates, are explained, in our opinion, by frost weathering (Fig. 24).

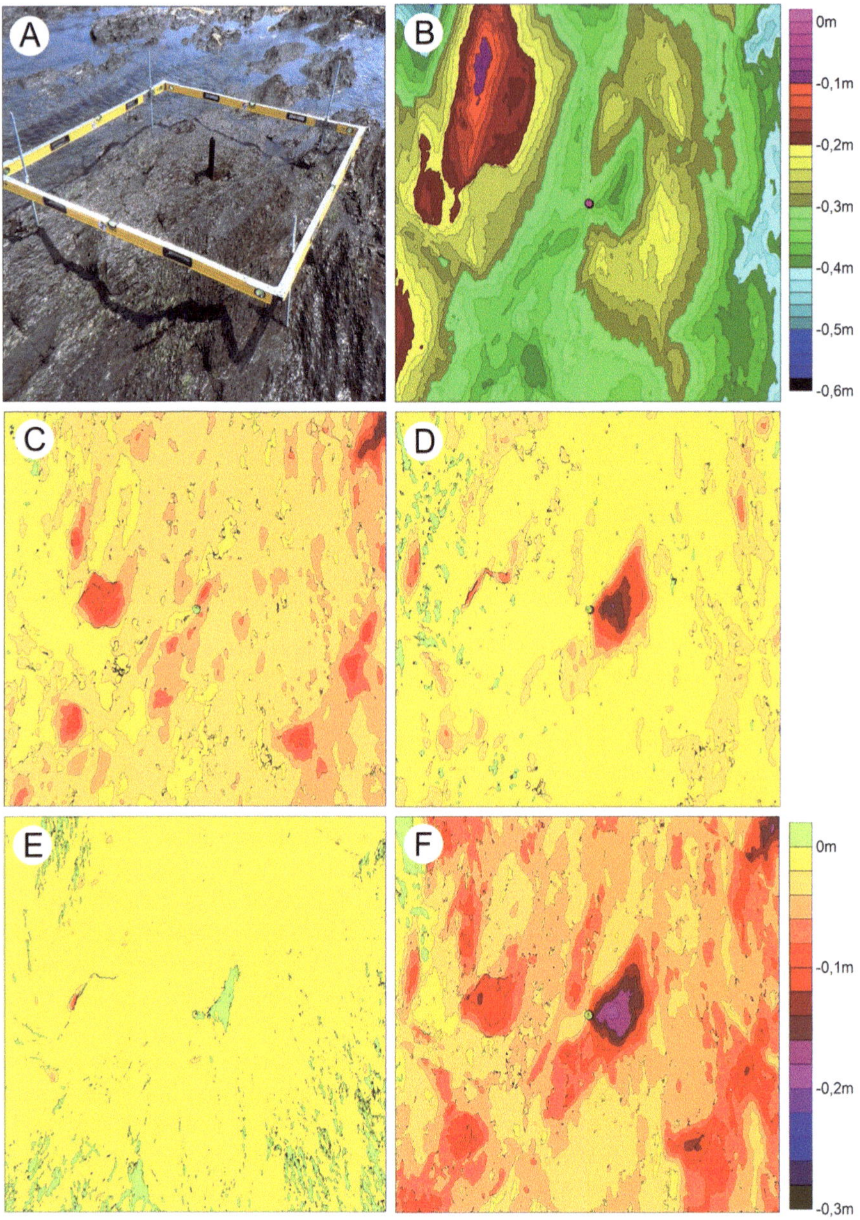

Fig. 15 Elevation field maps and elevation field differences by year (rep 5/2): (**a**) site photo, July 2016, (**b**) site elevation map, November 2018, (**c**) elevation differences, 2017.06-2015.07, (**d**) elevation differences, 2016.04-2017.06, (**e**) elevation differences, 2016.11-2016.04, (**f**) elevation differences, 2016.11-2015.07

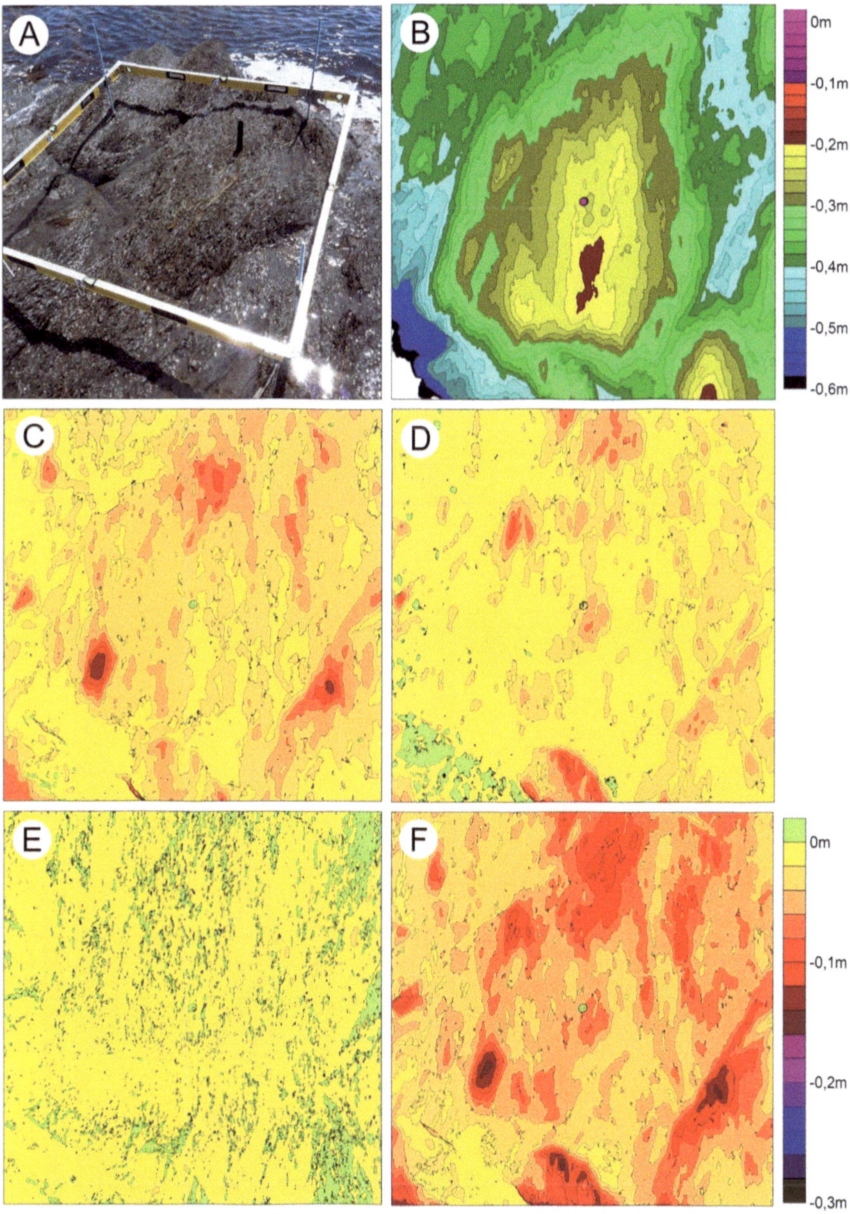

Fig. 16 Elevation field maps and elevation field differences by year (rep 7/2): (**a**) site photo, July 2016, (**b**) site elevation map, November 2018, (**c**) elevation differences, 2017.06-2015.07, (**d**) elevation differences, 2016.04-2017.06, (**e**) elevation differences, 2016.11-2016.04, (**f**) elevation differences, 2016.11-2015.07

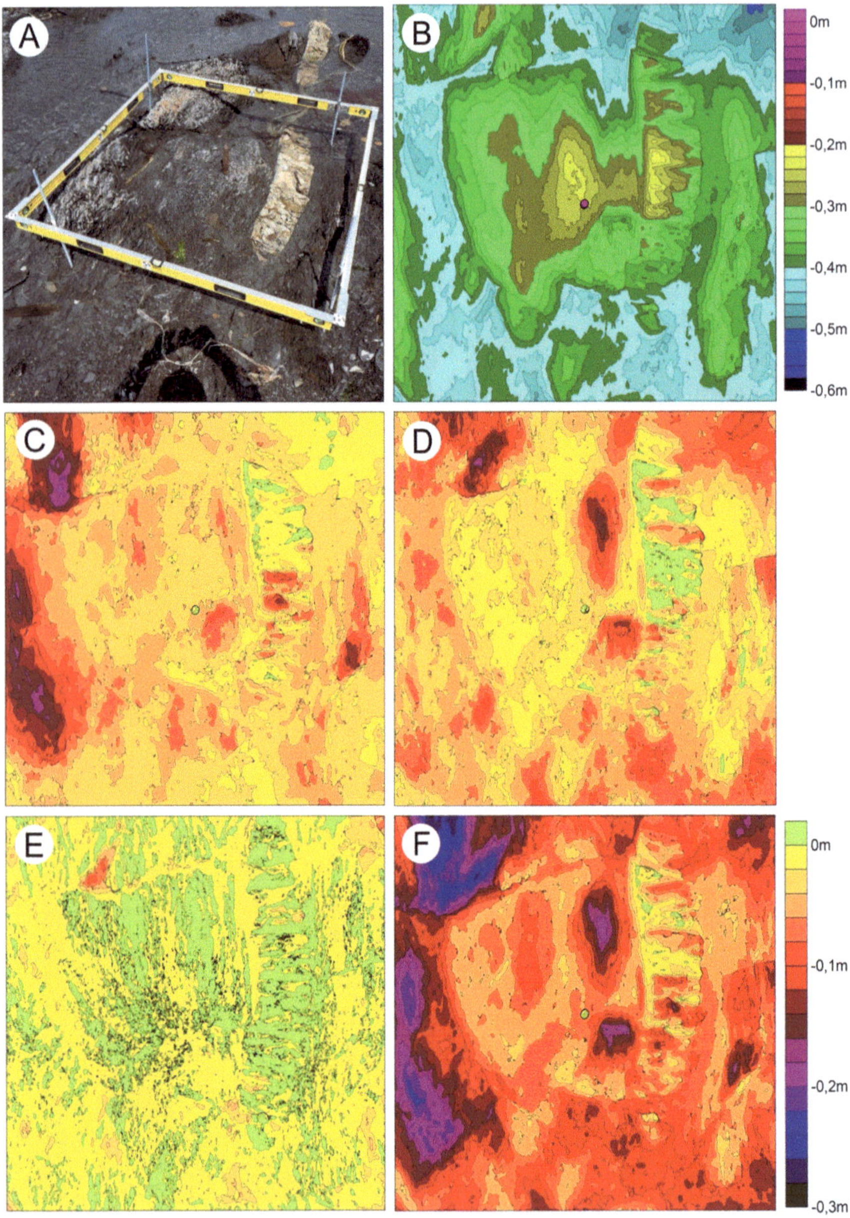

Fig. 17 Maps of height field and height field differences by years (rep 14/1): (**a**) site photo, July 2016, (**b**) site elevation map, November 2018, (**c**) elevation differences, 2017.06-2015.07, (**d**) elevation differences, 2016.04-2017.06, (**e**) elevation differences, 2016.11-2016.04, (**f**) elevation differences, 2016.11-2015.07

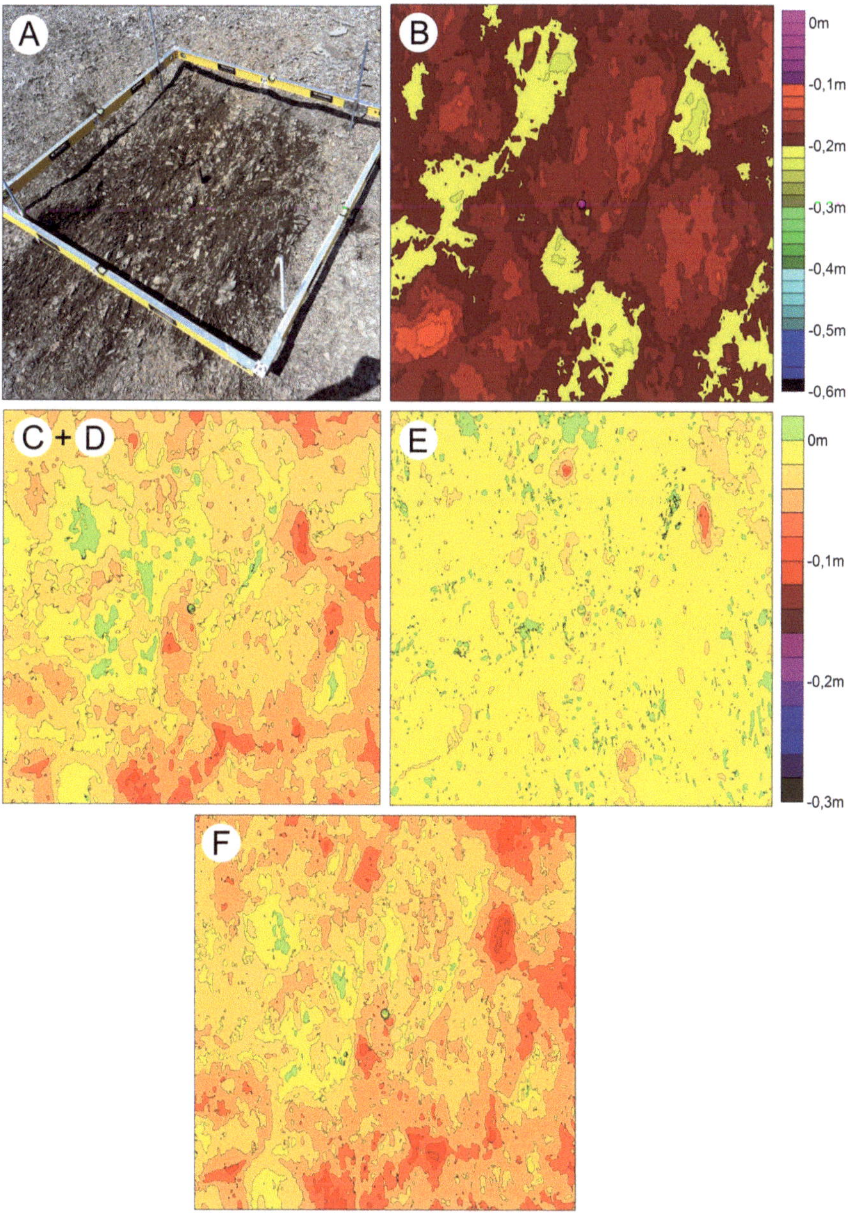

Fig. 18 Elevation field maps and elevation field differences by year (Rep (9/5): (**a**) site photo, July 2016, (**b**) site elevation map, November 2018, (**c**) elevation differences, 2017.06-2015.07, (**d**) elevation differences, 2016.04-2017.06, (**e**) elevation differences, 2016.11-2016.04, (**f**) elevation differences, 2016.11-2015.07

Fig. 19 Weathering products washed onto the inside of the benches

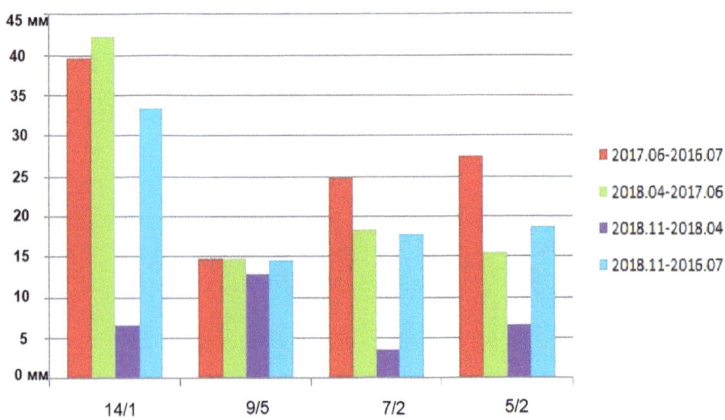

Fig. 20 Bench surface fracture rates at sites 14/1, 9/5, 7/2, 5/2

The mean annual duration of the period with mean daily soil temperature < 0 °C according to data of Hydrometeorological station Nevelsk is 138 days. Taking into account that landfast ice and solid ice cover are practically not observed in this area, the mean annual number of freezing and thawing cycles at subzero air temperatures during the open sea period in the bay part of the Bench is not less than 100 (Fig. 25, 26 and 27).

Fig. 21 Fracture of differently stable rocks for the period 2016–2016. Calcitised benches fragment is circled in red

Fig. 22 Reaper ¼

28.04.2012 г.

13.06.2013 г.

04.04.2017 г.

Thus, as a result of this study, it has been established that the destruction of subvertical coastal scarps in rocks of weak and medium stability under conditions of wave action occurs with velocities approximately one order of magnitude greater than the destruction velocities of subhorizontal surfaces of the benches composed of similar rocks. The measured retreat velocities of the coastal scarp without direct wave action are approximately the same as the failure rates of similarly stable rocks on the benches. In both cases, there are no shock wave loads with abrading effects

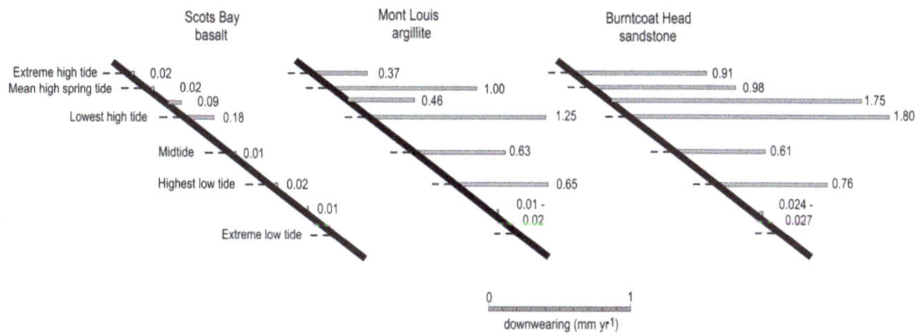

Fig. 23 Experimental fracture rates of rock samples obtained from shore platforms in eastern Canada (Porter 2010)

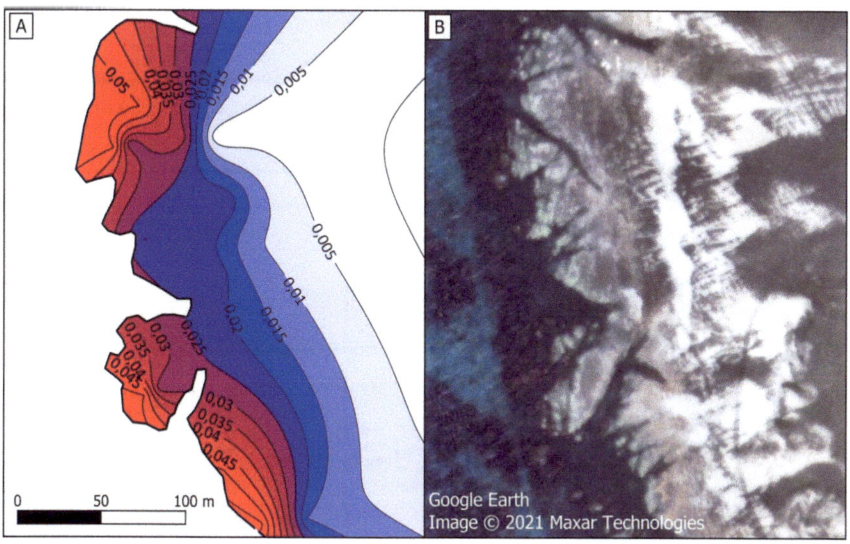

Fig. 24 Average annual fracture rates of the Bench surface; (**a**) a diagram of deformation (isolines are drawn at 0.005 m intervals), (**b**) a visualisation of the site

of beach sediments. And if from Bench the weathering products are removed by sea water, in the case of cliff it is due to slope, mainly gravitational processes.

Finally, we note that the longest study (43 years) of surface erosion of a mudstone-bedded benche in New Zealand using micro-erosion counters has shown that erosion rates from short-term observations correlate well with full-series data (Stephenson 2019). In 2016, the benches at a monitoring site in New Zealand were raised by 1.01 m (Stephenson 2017). Thus, the conditions of platform development changed and a number of observations were interrupted. However, observations have been continued under the new conditions and it is hoped that also, as in our case, will improve the understanding of coastal platform development processes (Afanas'ev 2013).

Fig. 25 Repeats 1, 2, 3 of Stream 12 (09.02.2018)

Fig. 26 Bench surface in winter period (22.01.2017)

Affordable, low-cost photogrammetric methods also allow us to obtain highly accurate data on high steep coastal escarpments composed of bedrock (Westoby 2012; Afanas'ev 2016). As an example, let us consider volcanogenic shores widely represented in the Far East (Fig. 28).

The detailed study area is located on the northern coast of Chikhachev Bay (De-Kastri settlement), northwestern part of the Tatar Strait within the Tumninsky neotectonic uplift of the East Sikhote-Alin volcanogenic complex (Map of

Fig. 27 Formation of a micro-beach at the edge of eroding ice at Bench (22.01.2017)

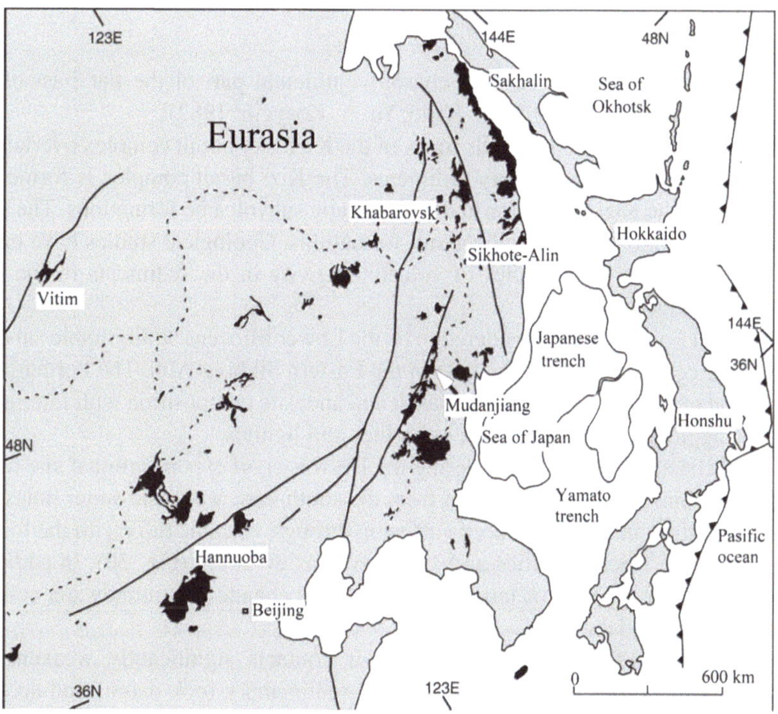

Fig. 28 Simplified map of Cenozoic volcanic fields of north-eastern Eurasia (Okamura 2004)

Fig. 29 Structure of the coastal scarp in Chikhachev Bay. Chikhachev Bay (orthophotoplane fragment)

volcano-tectonic structures of the coastal-continental part of the Far East of the USSR. 1:1500000 / Ch. ed. N. A. Shilo, Yu. A. Kosygin. 1982).

The coastal escarpment reveals rocks of the Kizinsky basalt complex overlain by a thin cover of loose Quaternary sediments. The Kizi basalt complex is formed by sediments of the Kizi Formation and komagmatic subvolcanic formations. The suite is subdivided into lower and upper sub-formations. Geological studies have established approximately 80 cycles of volcanic activity in the sediments of the Kizi Formation (Fig. 29).

The Kizi Formation is considered to be the Lower Miocene stratigraphic subdivision of the central and northern parts of the Eastern Sikhote-Alin. The Formation is composed of lavas of basalt, andesibasalt and andesite composition with interlayers of tuffs, agglomerates, mudstones, ophiolites and lignites.

In the coastal escarpment bordering the De-Kastri oil export terminal site of the Sakhalin Mainland Pipeline System from the south-east, we found water holes and erosion grottoes indicative of deep suffosion through volcanic tuffs with the formation of density heterogeneities and voids (Afanas'ev 2015) (Fig. 30). In addition, two observation wells in the terminal area showed changes in groundwater composition as part of tidal cycles.

Subsurface suffusion on tuffs and their contacts significantly weakens the strength characteristics of the volcanogenic-sedimentary rock massif and accelerates weathering processes on the slopes of the coastal escarpment. An express indicator of these dangerous geological phenomena was the presence of groundwater outlets with a high degree of mineralisation. In De-Kastri conditions the mineralisation of water outlets is 2–4 times higher than the mineralisation of water at the

Fig. 30 Coastal scarp with a wave-surge niche at the base. Grottos formed by groundwater outcrops are observed at 30–40 metres height

mouth of rivers. The area is located in a nine-point seismic active zone, and in case of seismic events the loads on the basement rocks will increase by orders of magnitude.

The above considerations allowed us to investigate a more than 5 km long coastal escarpment bounding from the south and east the site of the planned liquefied natural gas (LNG) plant construction, located next to the oil export terminal near the Klykova Peninsula (De-Kastri).

The current geological and geomorphological appearance of the site is determined by the formation of a series of effusive formations in the Miocene and their subsequent exogenous reworking. Over the last 6 thousand years, a benches up to 100 m wide have been formed as a result of the destruction of the coastal scarp up to 160 m high.

Groundwater by the type of filtration medium is formation-cracked in the upper, most fractured zone and fracture-vein in the zones of tectonic disturbances. Both aquifers described above in this area have a close hydraulic relationship and are characterised by a common level surface. In general, the depth of the level decreases in accordance with the lowering of the topography, and the maximum gradient is observed as the coastal escarpment is approached.

Orthophotomaps of the coastal scarp were constructed in Agisoft Photoscan software by projecting a set of ground photos taken with a SONY NEX camera onto vertical planes parallel to the scarp (User's Guide Agisoft PhotoScan Professional Edition, version 1.2 [Text] / 2015.—116 p). Photogrammetric reconstruction of the 3D model of the ledge was performed to obtain a correct orthophoto projection. For the most accurate matching, all photos were taken in RAW format with subsequent conversion to TIFF with extended dynamic range. The average resolution of the photos was 2.64 cm/pix. Initial calibration of the camera was carried out by in-built tools of Agisoft Lens software. Then photo linking was performed and coordinates of photo centres were calculated by constructing a sparse cloud. A total of five million linking points were calculated. The lens distortion model was refined using recognised connecting points, which were registered in at least 3 images, and then a dense point cloud was built, totalling 62 million points. From the dense point cloud, a polygonal model was generated and used to produce orthophotos. The model was georeferenced by coordinated markers located on the ledge. The total length of the coastal scarp studied exceeds 5 km.

Directly within the proposed activity site, 7 water occurrences in the form of wetted spots several metres in size were detected during the analysis of models and orthophotos. Outside the proposed activity site during the observation period groundwater discharge was carried out both through contact fracture zones and tuffs. More than 30 zones with water seepage were identified along the entire height of the bank scarp. At the same time, no water occurrences were identified by route observations along the foot of the bank scarp.

The application of monitoring techniques using remote sensing data and observation stations is not possible on high steep cliffs due to the sloping viewing angles, heterogeneous structure and complex scarp surface. Repeated photogrammetric surveys in a fixed coordinate system allow to solve monitoring tasks with accuracy not inferior to laser scanning technologies (Bemis 2014).

Nevertheless, at the first stage of the study of high coastal scarps, we assessed the rates of shore failure, especially when shore scarp areas with easily eroded geological complexes or disintegrated complexes of disjunctive zones enter the surf zone. The morphometric parameters of the coastal scarp in the areas of active slope processes provoked by abrasion undercutting of the scarp or reduction of rock stability as a result of weathering and suffosion were determined, and the volumes of colluvial-proluvial forms at the foot of the scarp were determined (Fig. 31).

The time of formation of colluvial-proluvial formations and their elements was established on the basis of analysis of remote sensing materials and the nature of vegetation on the surface. The averaged rates of bank scarp destruction by zones of accelerated destruction are more than 0.29 m/year in the first zone, more than 0.44 m/year in the second zone, and more than 0.87 m/year in the third zone. One-time retreat of the coastal scarp in areas up to 100 m long can reach 7 m per year, which is somewhat unusual for bedrock.

It should be noted that the rockfall material disintegrates rather quickly in the surf zone. Large (up to 10 metres) fragments of tuffs and tuffobreccia are literally crushed by large pebbles and boulders of basalt varieties. Pebbles of tuff varieties are practically absent in the beach sediments, despite the fact that more than 60% of the total volume of incoming material falls into the surf zone during the destruction of the tuff ledge.

Tuffs are exposed on a significant part of the coast in the nearshore zone, in which wave-surge niches are formed quite quickly, provoking the processes of subsidence and destruction of the coastal scarp (Fig. 32).

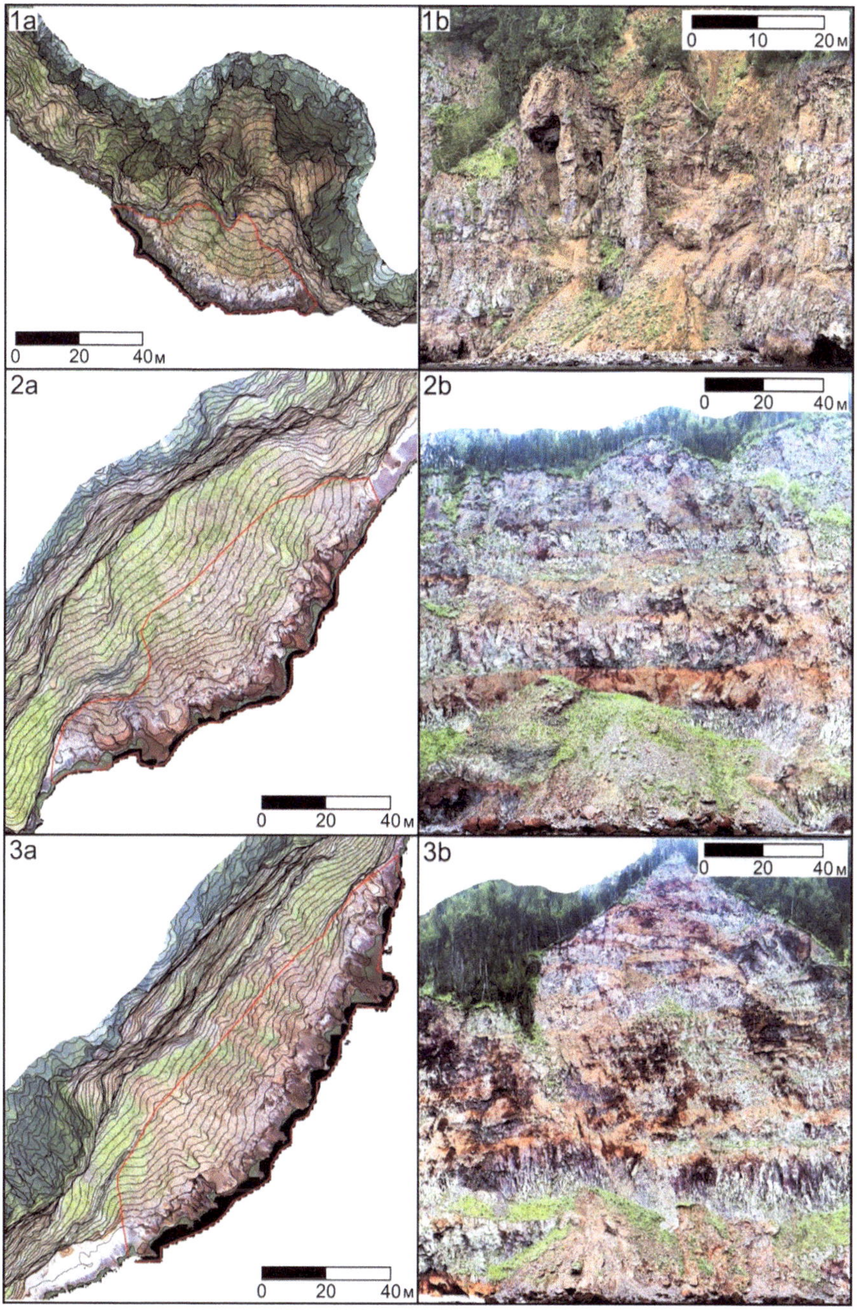

Fig. 31 Coastal scarp at three sites of active slope processes: (**a**) colluvial-proluvial formations at the foot of the scarp; (**b**) orthophotos of the coastal scarp

Fig. 32 Fracture subsidence and tuff collapse

It should also be noted that geotechnical drilling to bedrock, in the case of interbedded monolithic andesite-basalt and "weak" tuffs, makes no sense, because the thickness of the "bedrock", reached after sinking of slope deposits and eluvium, can reach only 1–2 m.

3 Scouring of Banks Composed of Unconsolidated Neogene-Quaternary Rocks

Sakhalin Island is one of the few regions in Russia where the parameters of modern erosion of shores composed of loose and poorly consolidated Neopleistocene-Holocene sediments have been determined. Examination of cartographic information and geological descriptions of Sakhalin has shown that the observed spatial pattern of erosion of marine terraces and embankments has existed in basic terms since at least the 1930s (Afanas'ev 1991). In analysing the spatial pattern of erosion of the low marine Holocene-Pleistocene plains of southern and north-western Sakhalin, erosion rhythms with wavelengths of 4–5 km, 11–12 km, ≈30 km and ≈60 km were identified (Fig. 33).

The main patterns of spatial patterns of coastal erosion were established on the basis of analyses of aerial photographs (APS) for the period 1952-1980-2005. Measurements were made at 100, 200-metre intervals. The accuracy of measurements is 1.5–2.5 metres. If this error is correlated with the observation period, the error in determining the mean annual bank retreat rate will be 5–10 cm.

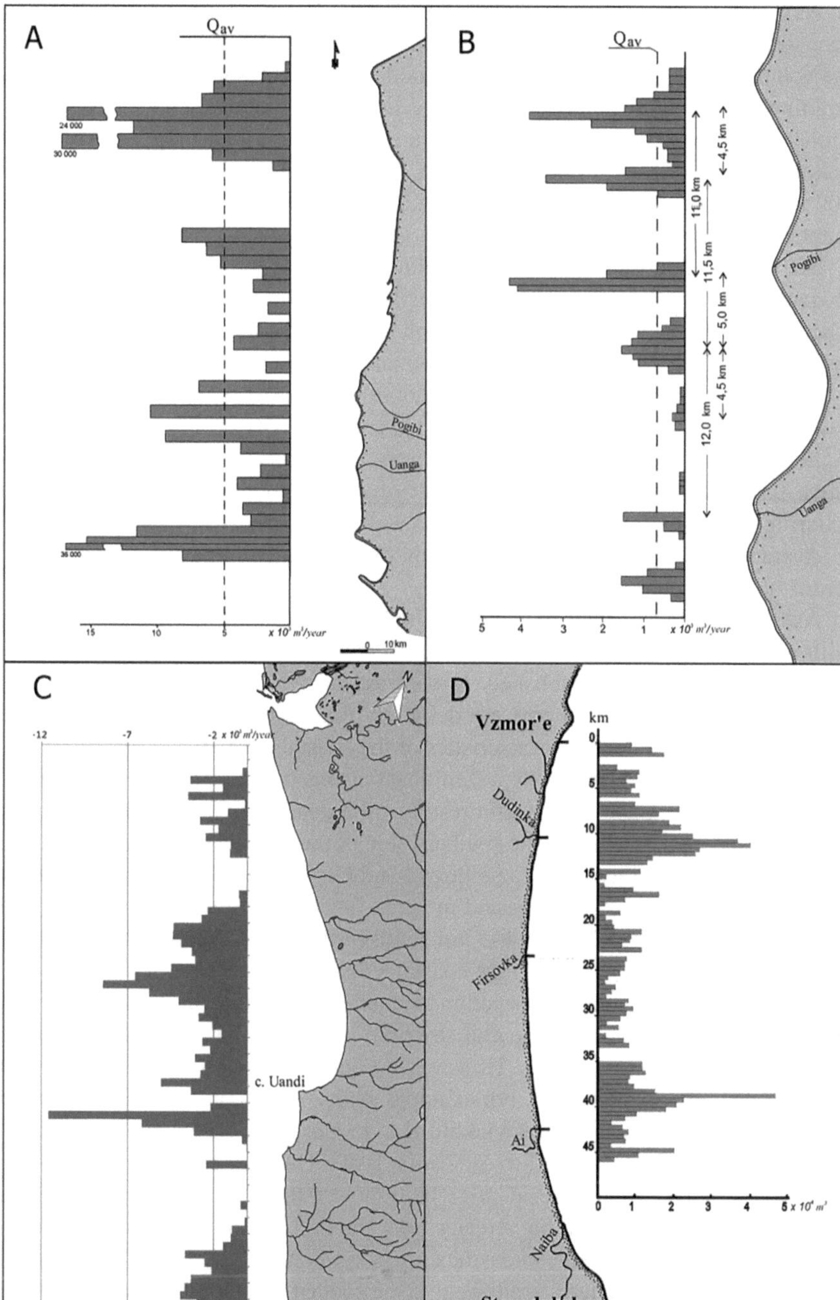

Fig. 33 Spatial plans of coastal scour: (**a**) north-west Sakhalin (averaged over 5 km sections over a period of 30 years), (**b**) Nevelsky Strait (averaged over 500 m sections over a period of 30 years), (**c**) Tatar Strait (averaged over 500 m sections over a period of 40 years), (**d**) Terpeniya Bay (averaged over 500 m sections over a period of 40 years)

As a consequence, areas of the coastal scarp eroding at a rate of less than 0.2 m/year were categorised as stable and were not taken into account when calculating the volume of erosion. Road segments, wooden and metal supports of power lines, buildings and natural objects that change little over time (relief elements, lake-shores, individual trees) were used as reference points. The average annual scour volume of the north-west Sakhalin coastal scarp, which has a length of more than 300 km, was calculated for 5 km sections. This reduced the influence of measurement error and low-order variability in the intensity of coastal erosion (Fig. 33a).

The distribution of bank erosion volume values in north-western Sakhalin is characterised by periodicity with a wavelength of approximately 60 km. Interestingly, when analysing the intensity of abrasion on the open coast of the USA, a value for the wavelength of the scour rhythm was obtained that is approximately half of that found (Dolan 1981). The above pattern is not fulfilled in the area of "Azov" type spit development. Under present-day conditions, scour activity in this area "should be higher" than the coastal average. Nevertheless, the degradation of modern accumulative formations noted in the second half of the twentieth century indicates that scouring processes are intensifying here as well.

Averaging the volume of coastal erosion over 500-metre intervals reveals a different system of change in scour activity.

As an example, consider the coast of the Nevelsky Strait, where scour rhythms with wavelengths of 4–5 and 11–12 km and the area of the Azov-type spits with a 1.5–2.5 km rhythm are established. In contrast to high-order, in our case 60 km, rhythmicity, these scour patterns are definitely related to the configuration of the coastline and wave-energy characteristics of the coastal zone.

It should be noted that despite a significant change in the exogeodynamic situation in the area of Cape Pogibi as a result of the construction of a one and a half kilometer sheet piling dam on the southern side of the cape in 1952, the volumes of erosion of the coastal ledge of Cape Pogibi and Cape Tungusky are approximately equal and amount to 11–12 thousand m^3/year. It seems interesting that there is an area of erosion in areas located 4–5 km south of the capes. About 7 thousand m^3/year enters the coastal zone from the coastal ledges here.

Analysis of cartographic information available for the study area has allowed us to establish that the observed spatial structure of erosion has existed in its basic features since at least the 1920s. However, coastal erosion in north-west Sakhalin has occurred on a smaller scale. For example, between 1940 and 1980, in a section of approximately 100 km (from Viakhtu Bay to Uaz), the extent of eroded banks increased from 31.4% in 1940 to 50.2% in 1980. The data are confirmed by the results of the survey of local residents, which showed that up to the forties the banks were practically not eroded, but already from the mid-fifties significant rearrangements were noted. Despite the intensification of erosion processes in recent decades, a 3–4 metre long marine terrace adjacent to the sodden coastal escarpment has been fragmentarily preserved to date. Radiocarbon age determinations of woody remains (finwood) from three sections (Tungussky, Uanga and the mouth of the Malaya Pogibinka River) allow us to assign it to modern formations. The oldest larch trees growing on one of the terrace fragments in the Nevelsky Strait area are 60–80 years

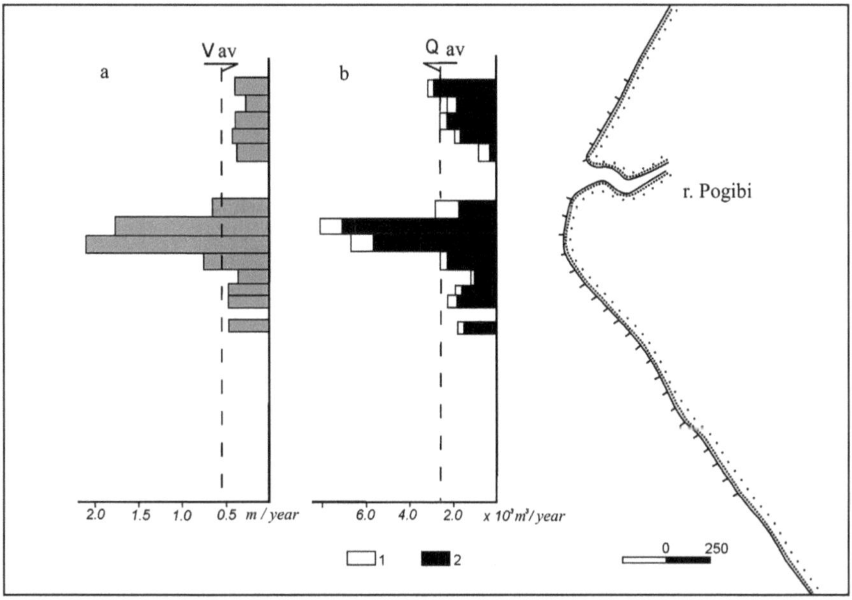

Fig. 34 Beach scour at Cape Pogibi; (**a**) areas, (**b**) volumes, 2—fraction of beach-forming sediments

old (9 determinations). Given the peculiarities of the morpholithodynamic regime of the coastal zone of the area, it can be assumed that the intensification of scouring processes and, consequently, of sediment flow rates between 200 and 300 years ago, which is confirmed by palaeogeographic information, led to a change in the spatial structure of abrasion and accumulation processes. In particular, in the Nevelsky Strait there was a shift in accumulation zones, accompanied in some cases by their expansion and in others by their degradation.

Considering abrasion as a factor that provides the largest revenue part of the budget, it is necessary to have an idea of the fractional composition of eroded sediments. Granulometric analysis of sediments composing the coastal scarp has shown that the content of beach-forming fractions (≥0.1 mm) varies from 0–15% (lake-swamp, lagoonal) to 70–90% (alluvial, coastal-marine sediments).

In the detailed studied coastal section of Cape Pogibi, the proportion of sand and gravel material along 100-metre intervals of the coastline varies from 30 to 90%. The average value for the 1400 metre section is 82% (Fig. 34).

In assessing the composition of sediment entering the water area during coastal erosion of the whole of north-west Sakhalin, the lower limit of silty-pelitic particles in coastal scarps of considerable length is approximately 20%, while for the coast as a whole the amount of suspended sediment slightly exceeds the amount of beach-forming sediment and is approximately 55–65% (Fig. 35). Thus, the sand and gravel yield from scouring of the coastal plain of north-west Sakhalin is significantly less than the maximum possible values (if only alluvial and coastal-marine sediments

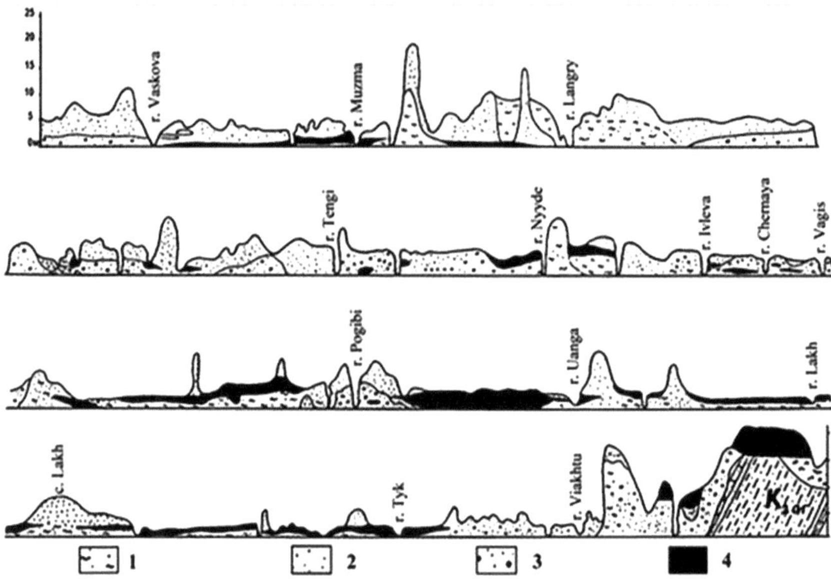

Fig. 35 Lithological section of the coastal escarpment of north-western Sakhalin: 1—silty-pelitic sediments, 2—multigrained sands, 3—gravel and pebble sediments, 4—peat (according to V.A. Konovalenko, 1974)

were scoured into the coastal escarpment) obtained by extrapolation of data on the fractional composition of the sediments of the Pogibi River. Pogibi for the whole area (2 × 10^5 m^3/year) and, even more so, the estimate of 2 × 10^6 m^3/year proposed by A.I. Solovyov when analysing the conditions of the Amur estuary fairway drift (Solov'ev 1974).

In assessing the composition of sediment entering the water area during coastal erosion of the whole of north-west Sakhalin, the lower limit of silty-pelitic particles in coastal scarps of considerable length is approximately 20%, while for the coast as a whole the amount of suspended sediment slightly exceeds the amount of beach-forming sediment and is approximately 55–65% (Fig. 35). Thus, the sand and gravel yield from scouring of the coastal plain of north-west Sakhalin is significantly less than the maximum possible values (if only alluvial and coastal-marine sediments were scoured into the coastal escarpment) obtained by extrapolation of data on the fractional composition of the sediments of the Pogibi River. Pogibi for the whole area (2 × 105 m^3/year) and, even more so, the estimate of 2 × 106 m^3/year proposed by A.I. Solovyov when analysing the conditions of the Amur estuary fairway drift (Solov'ev 1974).

Obtaining quantitative indicators when studying the morpholithodynamics of the coastal zone is not limited only to in situ observations, but involves the use of indirect, calculated methods, especially when analysing the factors of morpholithogenesis: waves, runup and surge level changes, tides, etc.

Obtaining quantitative indicators in the study of coastal morpholithodynamics is not limited only to in situ observations, but also involves the use of indirect, computational methods, especially when analysing the factors of morpholithogenesis: wave action, runup and surge level changes, tides, etc.

In accordance with the main theoretical provisions of coastal zone dynamics, the normal (Tn) and longshore (Trn) components of the wave energy flux in the surf zone are the main indicators of wave action on the shore (Popov 1977).

The possibilities of different ways of calculating wave energy indices are now more or less clear (Zenkovich 1989; Knaps 1979; Piskareva 1986; Selivanov 1985; Yurkevich 1987). Geomorphological conditions and hydrometeorological study of the area determined the choice of the calculation method developed by the Problem Research Laboratory for the Development of the North (Popov 1981, 1982). This methodology is best suited to the conditions of a shallow sandy shore and does not require long-term wave observations, as the calculation of wave energy fluxes is based on wind data. The methodology used, like other hydrometeorological calculation methods, does not give precise values for sediment flux capacity and bank erosion rates. Nevertheless, it makes it possible to determine areas of convergence (convergence) and divergence (divergence) of the wave energy flux, and to identify features of the spatial and temporal variability of wave action on the coastal zone.

For the calculation, wind data of the Viahtu, Rybnovsk and Pogibi meteorological stations for the period 1961–1987 were used. The nature of the stations' representativeness (i.e., identification of the coastal areas where their data characterise the wind regime) was determined by analysing the vane openness roses. The data of Viahtu station thus reflect the wind regime from m. Wandi to m. Lakh; Pogibi station characterises the area from m. Lakh to m. Chingai, and Rybnovsk station—the northern coastal regions.

The main results of the calculations are presented in the form of plots of the multi-year course of calculated wind energy characteristics (Tn, Trn, Trasm. migr.) at the sites with maximum, minimum and average values for the area and a map at a scale of 1:1,000,000, which reflects the spatial variability of the average annual impact on the coastal zone (Figs. 36 and 37). Analyses of the cartographic and morphometric characteristics of the coastal zone and hydrodynamic parameters of the tidal field formed the basis of a meaningful model of coastal scour in north-western Sakhalin. The model was then converted into statistical models to assess the closeness of relationships and to determine the level of significance. Multivariate correlation and regression analysis was used as a tool. All information was averaged over 5 km intervals. The series of function values consists of 38 elements.

The following characteristics of the coastal zone were taken as arguments:

x1—length of the coastal scarp;

x2—length of the coastline eroding at a rate of more than 0.2 m/year;

x3—slopes of the underwater coastal slope;

x4—drainage area;

x5—solid runoff (nannofossil sediments);

x6—beach area;

x7—mean annual Tn component of wave energy flux;

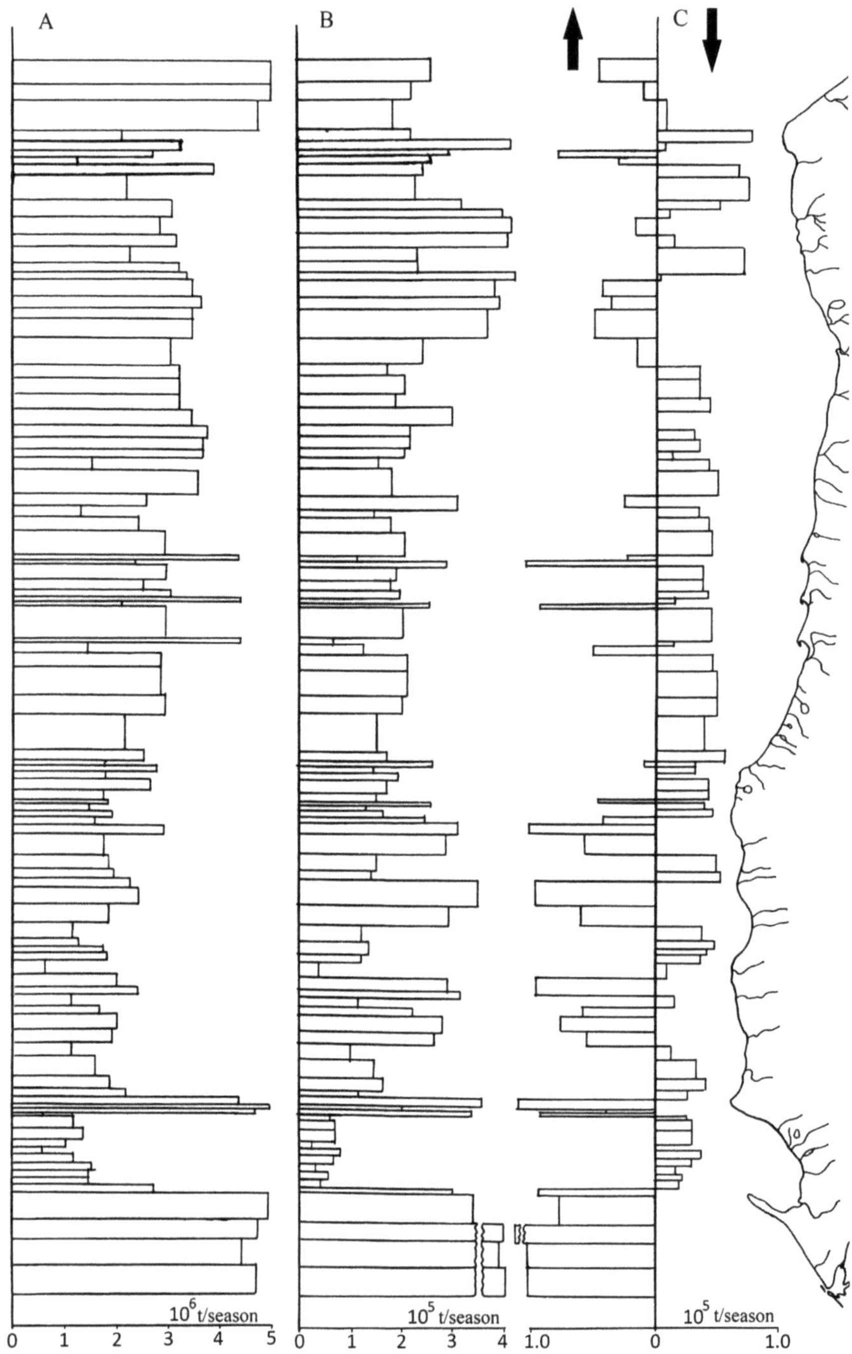

Fig. 36 Distribution of average annual calculated wind energy parameters along the coastline: (**a**) normal component of the wave energy flux; (**b**) migration range; (**c**) along shore component of the wave energy flux

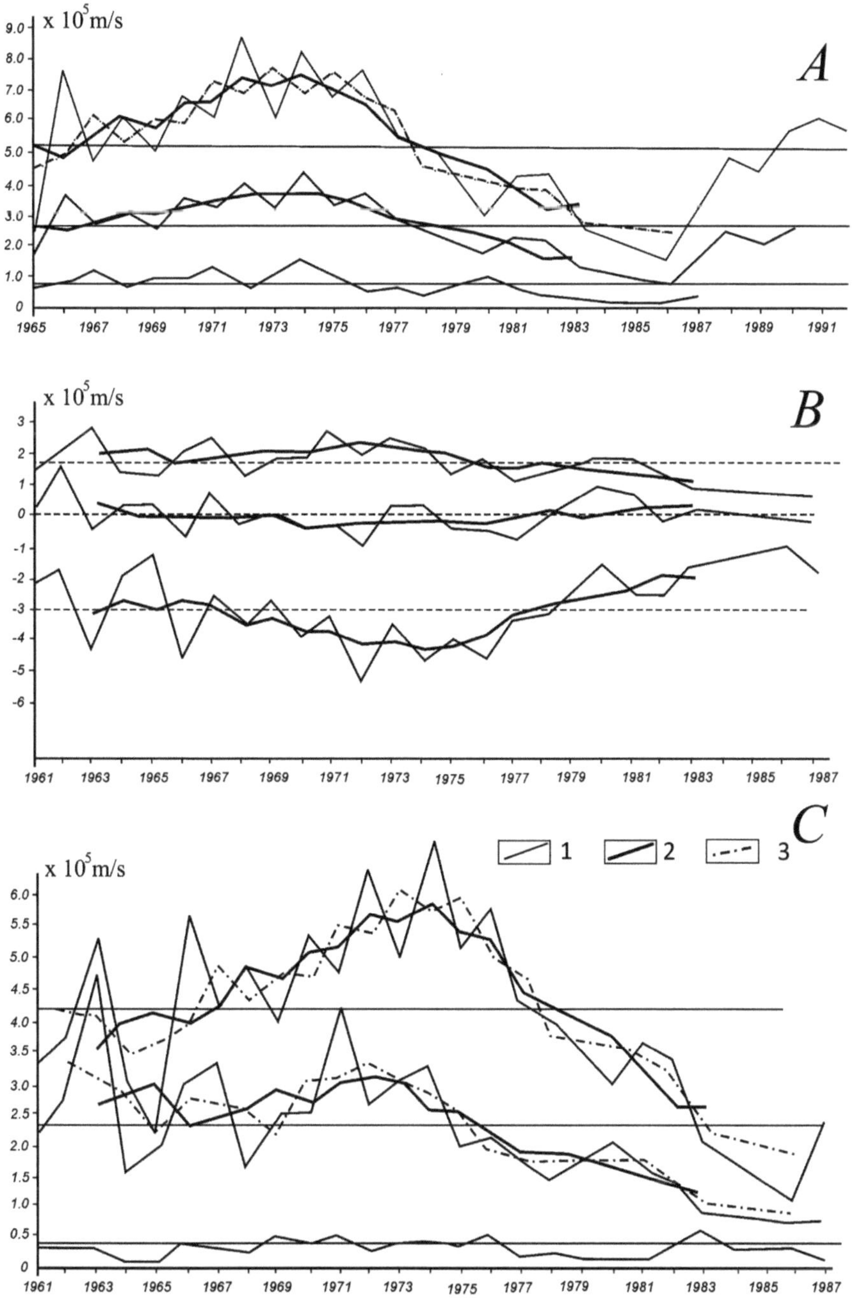

Fig. 37 Variation of wind energy characteristics of the coastal zone of north-western Sakhalin for sites with maximum, average and minimum values for the area: (**a**) Tp; (**b**) Tgp; (**c**) migration range; 1—calculated values; 2—averaged over 3 years; 3—averaged over 5 years

x8—mean annual resultant Trn component of the wave energy flux;

x9—mean annual total Trn component of the wave energy flux.

Taking into account that the coastal scarp is eroded mainly during storm waves, along with mean annual values, data on extreme storm years were entered into the matrix: x10 –n, x11 –Trn res., x12 –Trn total.

In the first step, pairwise relationships were investigated. When paired relationships were analysed, it was found that the most significant factors for coastal scour were submarine coastal slope gradients, beach area and Trn total. Very low level of relationships were observed for scour volumes and Tn of wave energy flux, solid runoff and length of shore scarp. It should be noted that coastal scarp extent, unlike scour volumes, has a significant level of association only with Tn, with the nature of the relationship being inverse and the values of extreme storm years being favoured. When considering pairwise relationships between other characteristics of the coastal zone, it turned out that the beach parameters determined from APS interpretation data of 1953, 1973, 1977, 1980 and instrumental observations in 1983–1988 are well correlated with the area of tidal drifts (inverse relationship), slopes of the underwater coastal slope (direct relationship), and the total value of the longshore wave energy flux (direct relationship). It should also be noted that the lowest level of correlation with other elements of the matrix has values of solid runoff. Thus, the correlation matrix gives a fairly complete picture of the nature and closeness of the relationships of the substantive model of bank scour.

At the next stage, a model of the relationship between the intensity of coastal scour in north-western Sakhalin and morphometric and morphodynamic characteristics of the coastal zone was constructed in the form of regression equations.

The linear regression equations obtained for scour intensity (Q) are as follows:

$$\text{Variant 1 at } R = 0.8406$$

$$\begin{aligned} Q_{theory} = &-6045,80 + 1603,96\,x_1 + 1809,59\,x_2 - 313,47\,x_3 + 317,16\,x_4 \\ &- 0,15\,x_5 + 135,48\,x_6 - 731,60\,x_7 - 2457,52\,x_8 + 2525,29\,x_9 \\ &+ 1655,47\,x_{10} - 1033,81\,x_{11} - 608,33\,x_{12} \end{aligned}$$

$$\text{Variant 2 at } R = 0,7471$$

$$Q_{theory} = 7565,65 + 652,80x_1 + 1775,72x_2 - 334,65x_3 - 109,66x_4$$

$$\text{Variant 3 at } R = 0,7488$$

$$Q_{theory} = 7730,04 + 628,85x_1 + 1761,52x_2 - 331,12x_3 - 100,11x_4 - 1,49x_5$$

$$\text{Variant 4 at } R = 0,7692$$

$$\begin{aligned} Q_{theory} = &\,748,73 + 570,33\,x_1 + 1824,84\,x_2 - 222,98\,x_3 + 125,62\,x_4 \\ &- 0,46\,x_5 + 192,48\,x_6 \end{aligned}$$

$$\text{Variant 5 at } R = 0{,}6489$$

$$Q_{theory} = 9939{,}98 - 465{,}59x_3 + 160{,}49x_4 - 1{,}90x_5 + 178{,}75x_6$$

$$\text{Variant 6 at } R = 0{,}6877$$

$$Q_{theory} = 10180{,}60 - 372{,}09x_3 + 19{,}15x_4 - 2{,}53x_5 + 78{,}20x_6 - 1281{,}05x_7 - 1600{,}23x_8 + 2396x_9$$

$$\text{Variant 7 at } R = 0{,}7337$$

$$Q_{theory} = 16046{,}5 - 578{,}53x_3 + 41{,}98x_4 - 3{,}47x_5 + 72{,}19x_6 - 853{,}84x_7 + 308{,}41x_8 + 2538{,}18x_9 + 565{,}38x_{10} - 2002{,}84x_{11} - 282{,}81x_{12}$$

$$\text{Variant 8 at } R = 0{,}5947$$

$$Q_{theory} = -1662{,}59 - 1605{,}94x_7 - 1264{,}77x_8 + 4263{,}69x_9$$

$$\text{Variant 9 at } R = 0{,}6106$$

$$Q_{theory} = -614{,}54 - 3654{,}63x_7 - 1240{,}54x_8 + 1826{,}31x_9 + 1087{,}64x_{10} - 1954{,}87x_{11} + 1907{,}61x_{12}$$

The equations of options 1 and 7 can be used as predictive equations. Analysis of option 1 shows that scour activity increases with increasing morphometric parameters such as drying width, beach width and submarine slope and decreases with increasing solid runoff. Beach width and solid runoff are the least significant variables in the equation. It follows from the same equation that the volume of scour is greater the higher the mean annual values of the migration spread, the values of the normal component of the wave energy flux in storm years, and the smaller the mean annual values of Tn and Trn, as well as Trn and Trn res. in extreme storm years. The significance of the x_{12} and x_8 arguments is an order of magnitude lower than the other variables included in the equation.

Variation 7, obtained without taking into account characteristics such as the extent of the eroding bank and bank scarp, is in general not much different from the one discussed above. Only the mean annual Trn value enters the equation with the opposite sign (+). The most significant variables in this equation are x_4, x_6, x_{12}.

Taking into account the special character of morphodynamic informativity of the spatial distribution of shore escarpments, we also evaluated the nature of the relationship between the length of the shore escarpment (L) and the observed morphometric and calculated hydrodynamic parameters. The following equations were obtained:

$$\text{Variant 1 at } R = 0,7699$$

$$L_{theory}. = 6,21 - 0,11x_2 - 0,03x_3 - 0,11x_4 - 0,001x_5 + 0,018x_6 + 0,51x_7 + 1,11x_8 - 0,20x_9 - 0,94x_{10} + 0,005x_{11} - 0,12x_{12}$$

$$\text{Variant 2 at } R = 0,1440$$

$$L_{theory}. = 1,25 + 0,96x_2 + 0,24x_3 - 0,047x_4$$

$$\text{Variant 3 at } R = 0,1924$$

$$L_{theory}. = 1,35 + 0,085x_2 + 0,026x_3 - 0,040x_4 - 0,001x_5$$

$$\text{Variant 4 at } R = 0,2109$$

$$L_{theory}. = 0,21 + 0,092x_2 + 0,040x_3 - 0,010x_4 - 0,009x_5 + 0,026x_6$$

$$\text{Variant 5 at } R = 0,1991$$

$$L_{theory}. = 0,73 + 0,027x_3 - 0,008x_4 - 0,001x_5 + 0,024x_6$$

$$\text{Variant 6 at } R = 0,6765$$

$$L_{theory}. = 3,53 + 0,040x_3 - 0,108x_4 - 0,0004x_5 + 0,018x_6 - 1,23x_7 + 1,18x_8 - 0,069x_9$$

$$\text{Variant 7 at } R = 0,7660$$

$$L_{theory}. = 5,37 - 0,019x_3 - 0,103x_4 - 0,001x_5 + 0,024x_6 + 0,64x_7 + 1,04x_8 - 0,23x_9 - 0,97x_{10} + 0,072x_{11} - 0,161x_{12}$$

$$\text{Variant 8 at } R = 0,5709$$

$$L_{theory}. = 3,87 - 0,93x_7 + 1,03x_8 - 0,11x_9$$

$$\text{Variant 9 at } R = 0,6363$$

$$L_{theory}. = 3,52 + 0,66x_7 + 0,87x_8 + 0,56x_9 - 0,52x_{10} + 0,11x_{11} - 0,39x_{12}$$

The equations of variants 1 and 7 can be used as predictive ones. It should be noted very low values of correlation coefficients of those equations that include only morphometric indicators as variables. Analysis of equations with high correlation coefficients shows that the length of coastal scarps decreases with increasing drainage area and solid runoff, and increases with increasing steepness of the submarine coastal slope and beach width. The most significant regression variables are x_4 and

x_5, and the least significant are x5. From the group of calculated variables the most significant are x_8 and x_{10}. At the same time, an increase in the length of the coastal scarp is observed with an increase in the mean annual values of Tn and Trn, Trn in extreme years and a decrease in Tn in anomalously stormy years and Trn total.

Summarising the above, we note that Tn and solid runoff have almost no effect on scour activity, but the gradients of the submarine coastal slope, Trn res, have a significant effect. The low level of relationships with the observed shore zone parameters can probably be explained by the formation of the observed distribution in other morpholithodynamic conditions.

The variability of the considered characteristics of the shore zone does not exhaust the variability of scour activity and shore scarp length for these samples, as the influence of unaccounted factors on scour processes is quite significant. The model of interrelation between measured and calculated characteristics of the coastal zone and abrasion activity obtained in the form of multiple regression equations corresponds quite well to the observation materials.

The coast of Terpeniya Bay began to intensively erode. The coast of Terpeniya Bay began to erode intensively from the early 1960s, but the areas with mean annual velocities of 1.8–2.2 m/year were eroded already in the early 1950s (Fig. 38).

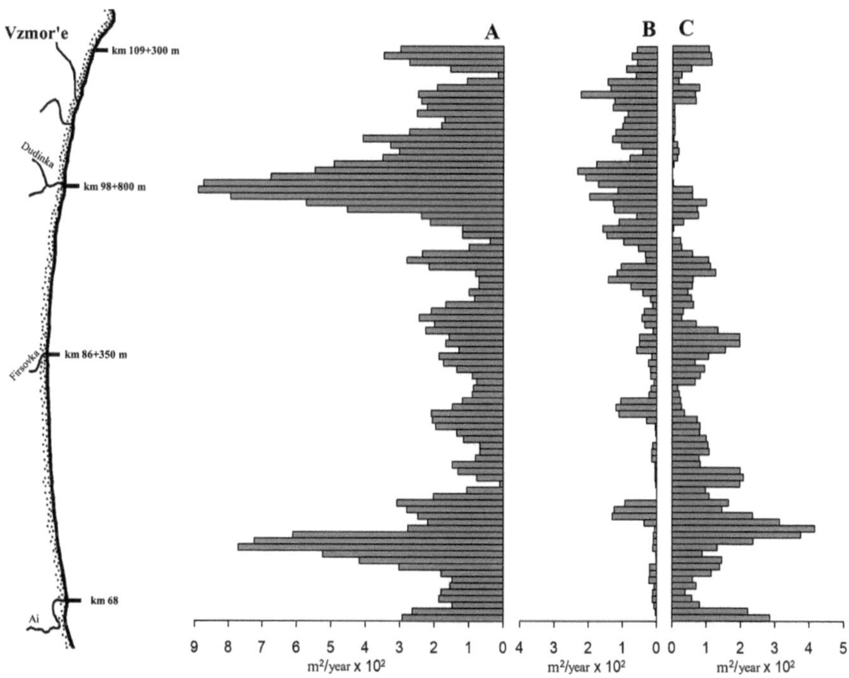

Fig. 38 Scouring of the southern shores of Terpeniya Bay at the Vzmorie-Starodubskoye site (mean annual values averaged over 500-metre intervals): (**a**) 1952–1991, (**b**) 1991–2013, (**c**) formation of low modern terraces after quarry closure

The spatial structure of bank erosion in the coastal area from Vzmorie to Starodubskoye is determined by the erosion rhythm with a wavelength of about 30 km. The coastal scarp 3–6 km to the north of Sovetskoye settlement and in the vicinity of Dudino station is most actively eroded. The average annual rate of bank retreat here is 2 m/year, with an average value of 0.5 m/year for the entire area. As noted above, when analysing the intensity of abrasion on the U.S. open coast, the value of the scour rhythm wavelength equal to that established by us was obtained (Dolan 1981). However, while there is no clear explanation for this on the US coast, in our case one "peak" is confined to the location of the Dudin quarry (where more than four million m^3 of sand has been extracted from the beach over several decades). The southern peak is confined to the Sovetskoye quarry, located in the zone of divergence of wave energy flows of the lithodynamic system Starodubskoye-Vzmorie. However, after the final closure of the quarries, scour in the zone of flow divergence was replaced by accumulation, which is associated with a large volume of solid runoff from two large rivers Naiba and Ai at the border of two coastal subsystems (divergence zone).

The shores of the northern part of the Tatar Strait are composed of regressive Neopleistocene sediments overlain by thick (up to 7 metres) Holocene peat bogs (Fig. 39). The Late Pleistocene stratum is composed of multigrained sands with interlayers of loam, gravel and pebbles, with peat lenses up to 1 m thick. The age of woody debris from the base of one of these peat pits was determined to be 29,565 ± 859 calendar years (FEFU -188). The terrace is overlain by aeolian sediments in areas of active overland drift (Fig. 39).

Processing of the aerial and space photo data sets (2014 and 2015) for the northern part of the Tatar Strait and north-east Sakhalin was already carried out in the GIS environment (Fig. 40). After implementation of the transformation procedures, scour areas were determined at 100 metre intervals. Mean annual bank scour rates for a period of 60 years were obtained. The areas for which the values of coastal zone deformation are less than the image resolution are categorised as stable. The maximum mean annual scour rates are 1.0–1.7 at a bank scarp height of 10–18 metres. Scour rhythms were established: 4.3 km, 9.1 km, 32 km.

4 Scouring of Complex Coastal Barrier Forms

With the materials on the erosion of barrier forms of the lagoonal coast of north-eastern Sakhalin we conclude the presentation of our studies on the determination of the parameters of modern coastal erosion on Sakhalin Island (Fig. 41).

The first geomorphological descriptions of the lagoonal coast of north-eastern Sakhalin are presented in the work of Tikhonovich N.N. and Polevoy P.I. (Tikhonovich 1915). Already in this work, the complex nature of the coastal structure was noted.

However, the first truly complete and detailed study devoted to the geomorphological structure and dynamics of the shores of northeastern Sakhalin is rightly considered to be the work of A.T. Vladimirov (1961). In this work, along with a number

Fig. 39 Neopleistocene terrace overlain by Late Holocene eolian sediments up to 5 m thick in the northern part of the Tatar Strait.The author takes a sample for radiocarbon analysis. Photo by Y.A. Mikishin

of conclusions about the structure and dynamics of the coasts that have not lost their relevance to the present day, the structure of longshore sediment movements was established on the basis of morphological analysis of the structure of the mouths of lagoonal straits and the 1952 Aerial photography data. Based on the results of interpretation of the 1952–1986 APS, bank scour parameters were determined at several coastal sites (Buzlaev 1993; Remenets 1981). In addition, morphodynamic observations on the Chayvo Lagoon embankment enabled a group of researchers from Far Eastern State University to establish that within the sandy embankment proper, sedimentation exceeds scour by a factor of about 6 (Volodarskiy 1983).

Our research covers the entire coast of northeastern Sakhalin, the time interval of analysed morphodynamic changes has been significantly extended, and modern methods of remote data processing have been used.

Average annual scour rates do not exceed 2 m/year anywhere. The maximum scour rates (0.4–0.6 m/year) averaged over the entire length of the shoreline are recorded at sites № 3 and № 5, which are high (14–28 m) terrace levels composed of Pleistocene sediments (mainly unconsolidated, sandy-pelitic fractions). These same areas are also characterised by the maximum volumes of sediment entering the coastal zone each year when a one metre long bank scarp is eroded (5.6–7.7 m^3). Localised areas of scour are formed during anthropogenic intervention where the rate of retreat of the coastal scarp exceeds 10–15 m/year, such as during the construction of the landfall of the underwater pipeline at Site № 2.

The spatial pattern and intensity of scour has the following characteristics.

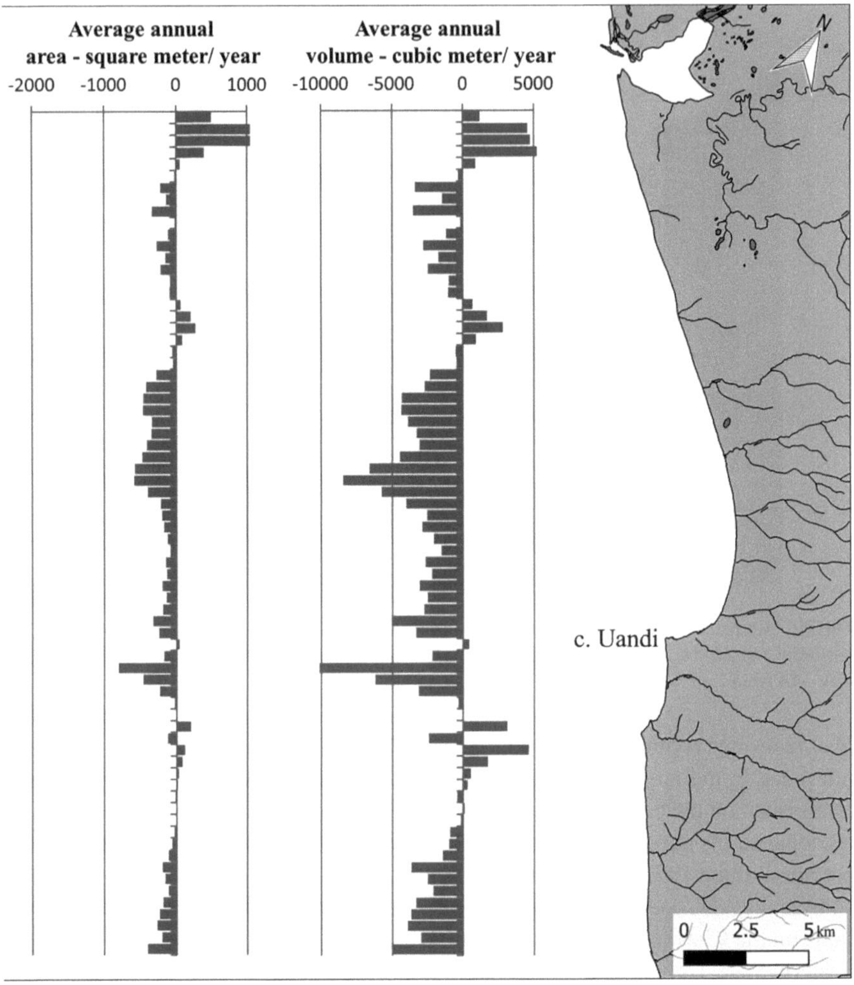

Fig. 40 Mean annual scour rates of Sakhalin shores in the northern part of the Tatar Strait (averaged over 500 metre intervals)

Three of the five major bays (essentially lagoons) and virtually all of the smaller lagoons of north-east Sakhalin are characterised by complexly constructed coastal barrier forms separating them from the Sea of Okhotsk. Areas of high coastal scarps composed of Pleistocene sediments alternate with Holocene overburden. There are no areas with high terrace levels in the zone of Late Holocene migration of the Straits. They were destroyed here during the first phases of coastal-marine accumulation (Afanas'ev 1992).

Data on scour intensity and activity were obtained by analysing aerial photoinformation (1952, 1974 flights) and space photoinformation (2014 and 2015). Processing of the data sets was carried out in a GIS environment. After the

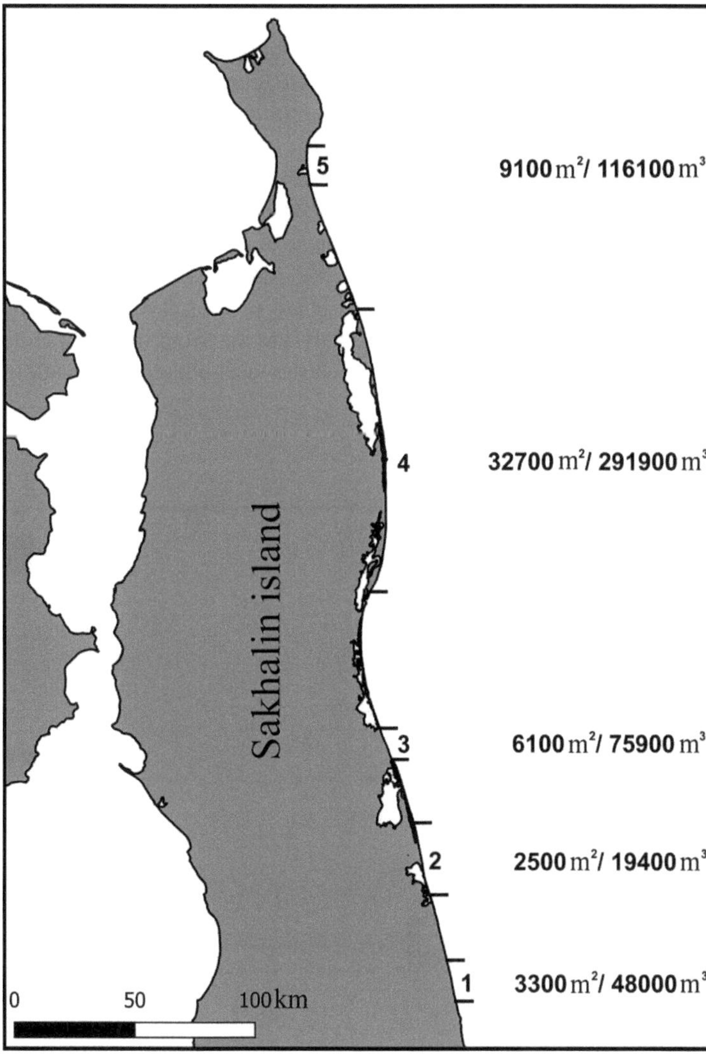

Fig. 41 Overview of the design sections. Plots: 1—Nampi-Kirinsky, 2—Lunsko-Nabilsky, 3—Nabil-Nyivensky, 4—Piltun-Chayvinsky, 5—Tropto-Shmidtovsky, 3300 m^2/48000 m^3—scour area per year/scour volume per year

implementation of transformation procedures, the areas of land loss were determined by hundred-metre intervals. As a result, the values of average annual bank scour rates for the period of 60 years were obtained. The areas with bank deformation values less than the image resolution were categorised as stable. In the next step, scour intensity spectral density plots were constructed by smoothing the periodogram with a Hamming window obtained from a fast Fourier transform (R.

The graphs show the contribution of scour processes of a certain periodicity to the total volume of bank failure, with the horizontal axis representing the spatial periodicity along the bank and the vertical axis representing the relative spectral power density, defined as the square of the amplitude of the variability of scour rate along the bank.

Figure 42 shows the mean annual scour areas at the maximum lengths of the Piltun-Chayvo (120 km) and Lunsko-Nabil (30 km) sections.

It should be noted that during the first phase of Holocene coastal-marine accumulation, the Piltun and Chaivo lagoons were connected. Separation of the reservoirs probably occurred during the second phase of coastal-marine sedimentation (Mikishin 2005; Afanas'ev 1992). When analysing the scour intensity of this complexly constructed Holocene-Pleistocene barrier accumulative formation with a

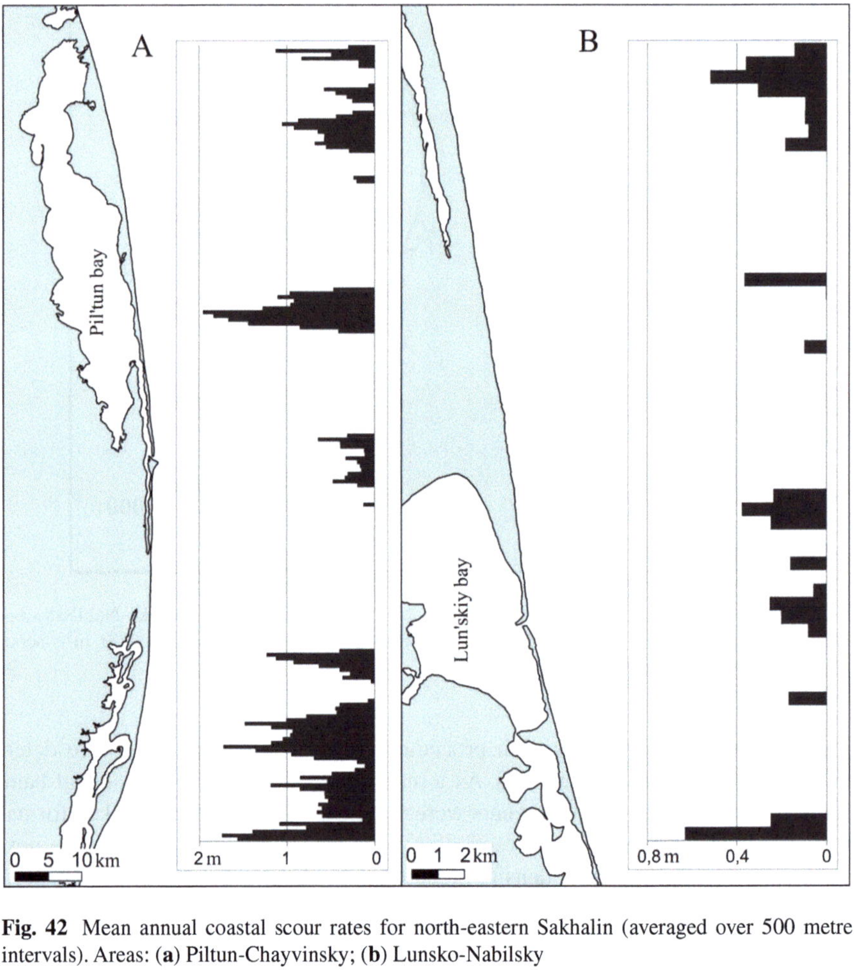

Fig. 42 Mean annual coastal scour rates for north-eastern Sakhalin (averaged over 500 metre intervals). Areas: (**a**) Piltun-Chayvinsky; (**b**) Lunsko-Nabilsky

length of about 120 km, values of the scour rhythm wavelengths of 4.0; 5.6; 6.5; 11.6; 25.6; 64 km were obtained (Fig. 43a).

These values are broadly consistent with the previously established spatial pattern of destruction of the low marine Holocene-Pleistocene low marine plains of southern and north-western Sakhalin.

5 Scouring of Accumulative Banks Formed in the Middle Late Holocene

The 30 km long Lunsko-Nabil section is a section of the coastline comprising the barrier formations of Lunskaya Lagoon, the southern eroded part of the Stary Nabil Hall embayment, and a mid-late Late Neolocene marine terrace about 4 km long

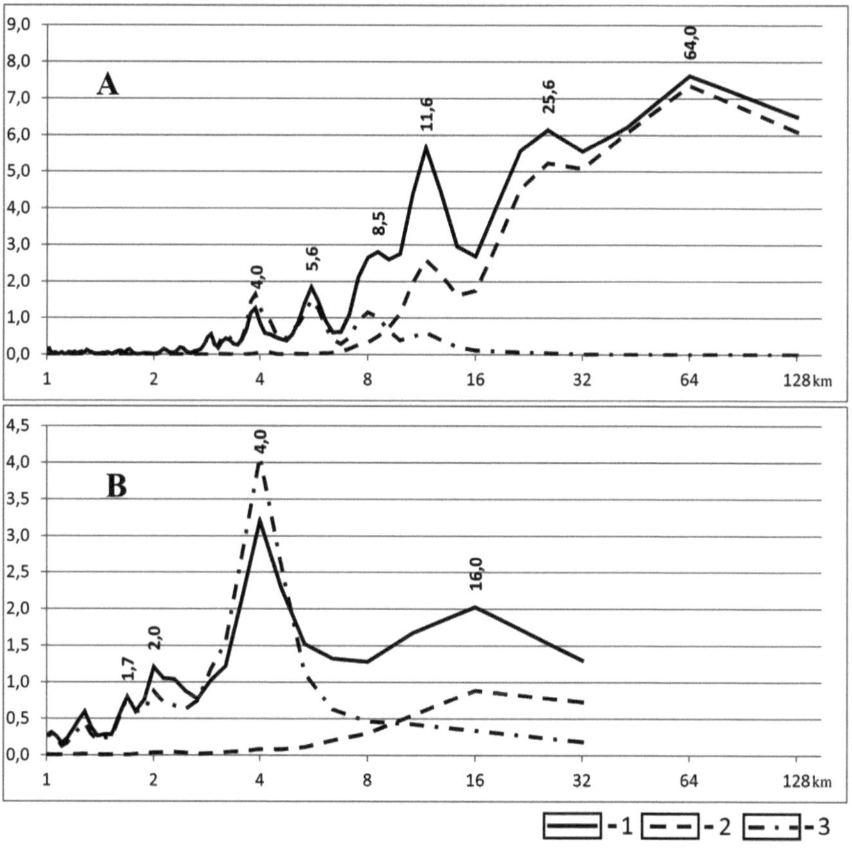

Fig. 43 Scour intensity spectral density plots. Plots: (**a**) Piltun—Chayvinsky; (**b**) Lunsko-Nabilsky; 1—initial data; 2—low-frequency; 3—high-frequency; X—periodicity along the shore; Y—relative spectral density

and 900–1100 m wide between them. Stary Nabil and a mid-late Late Holocene marine terrace about 4 km long and 900–1100 m wide between them. Almost the entire section of the terrace is avandune; the thickness of gravel and sand deposits in the zone of the coastal ramparts, according to drilling data, is 11–15 metres. The sediments of the wave field are underlain by aeolian deposits and heavy grey loams of lagoonal facies. The absolute mark of the roof of the lagoonal deposits in the avandune zone is −11 m (Baltic height system adopted in 1977). It is likely that the similar bank structure at the Lunsko-Nabilsky section caused the absence of a value in the 6.0–6.5 km interval on the spectral density plot of scour intensity (Fig. 43b). The indicated scour rhythm is characteristic of both the longest and the shortest scour plots (Fig. 44). The lowest order rhythmicity of scour, established by analysing the plots of spectral density of scour intensity, averaged over five hundred metre intervals, is 1.3 km. It is in this dimension that we have described the largest megafeston structures of the beach.

The insignificant contribution of scouring processes of this periodicity to the total volume of coastal erosion in the plots of the spectral density of scour intensity at ubiquitous propagation of sand waves and megafestons is explained by spatial and temporal peculiarities of scouring processes. Thus, only stable megafestons

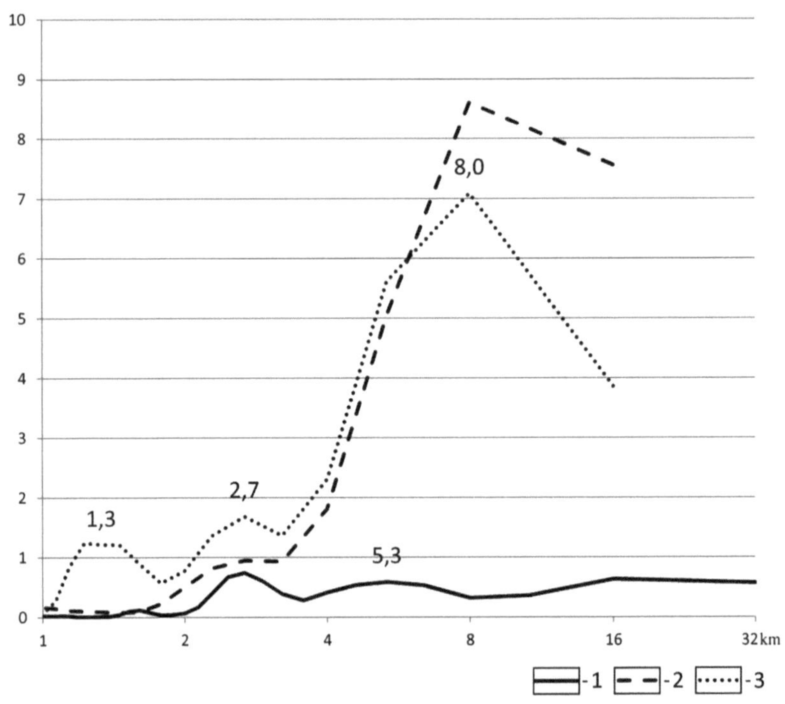

Fig. 44 Scour intensity spectral density plots. Plots: 1—Nampi-Kirinsky, 2—Nabil-Nyivo, 3—Tropto-Shmidtovsky, X—periodicity along the shore, Y—relative spectral density

similar to megafestons in the vicinity of Vzmorye were manifested in this time period of observations.

As a result of field observations, we found that all coastal types on Sakhalin Island are characterised by an uneven distribution of beach material along the shore—megafairstones and sand waves. The former are characteristic of open seashores with predominant zybe waves and infragravity waves, while the latter are characteristic of the Sakhalin coast of the Amur Liman, where only wind-driven waves are observed. These structures have a period from 200–300 to 1000–1400 m, in the zone of beach "capes" the beach width is 65–100 m. No erosion of terraces and avandunes is observed in these areas. The erosion zones are confined to the "concavities" of megafloods, where the beach width during the tidal phase is 3–8 m. The insignificant contribution of scour processes of this periodicity to the total volume of coastal erosion in the plots of spectral density of scour intensity at the ubiquitous propagation of sand waves and megafestons is explained by spatial and temporal peculiarities of scour processes. In the given time interval (60 years), only megafestons that exist steadily for a long time were manifested. An example of such a system in the vicinity of Vzmorie is given below. An example of such a system in the area of Vzmorye settlement is given below.

An interesting peculiarity of the development of a sandy accumulative form of the coastal barrier type—Plastun Spit—was established based on remote sensing materials for 80 years (1932–2013). During the period from 1932 to 1999, the spit grew northwards by 2300 m, while the sea side was practically not eroded. At the beginning of the twenty-first century, the growth process stopped, and we now observe accumulation in the root and distal parts of the spit, while the central part is eroding (Fig. 45).

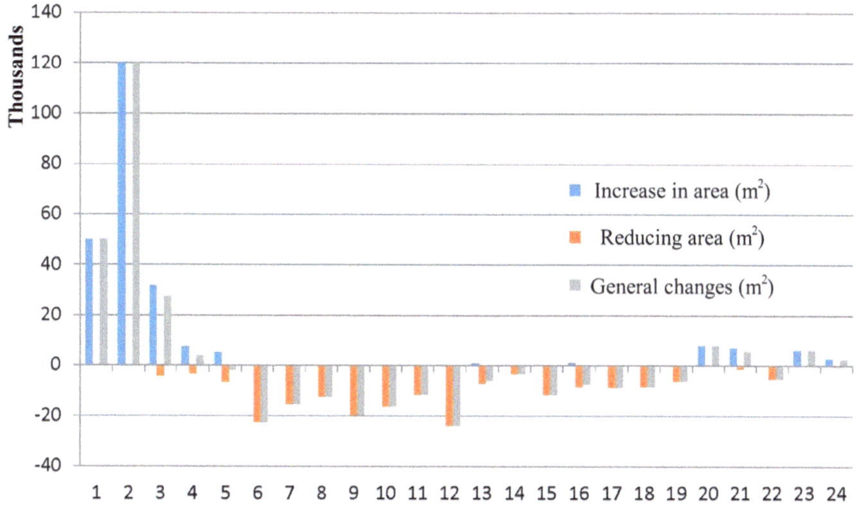

Fig. 45 Scour accumulation on the seaward side of Plastun Spit (averaged over 500 metre sections) for the period 1999–2014

Fig. 46 Seaward side of the Merea Lagoon embankment, anchored by rock revetment, with remnants of a wooden breakwater

It is interesting that during the periods of maximum growth of the spit (1952–1985) and stabilisation when this section was eroding, approximately the same amount of sediment was delivered to the coastal zone.

As scour has intensified since the 1930s–40s, the problem of protecting facilities and utilities on accumulative formations has become more acute. On Sakhalin Island, the remnants of Japanese concrete and wooden breakwalls that have survived on the coast are mainly located in areas where morpholithocomplexes of Middle and Late Holocene coastal-marine accumulation are virtually absent. In the post-war period, structures on the banks of low alluvial-marine terraces were added to the coastal protection system. However, over the nearly century-long history of coastal development on Sakhalin Island, only one barrier-type accretionary form has been constructed (Fig. 46).

On the one hand, this is due to the stability of inherited accumulative forms of the barrier type (Safyanov 1995). On the other hand, this is due to the uselessness of fixing the sea slope with rock revetments and breakwaters in conditions of eroding coastal bottom (Zenkovich 1987).

It should be noted that in the twentieth century coastal protection solutions on barrier accumulative forms in Okhotsk, western Kamchatka and Alaska proved ineffective (Fig. 47).

Nevertheless, in recent years the issue of protecting the Piltun and Chaivo lagoon embayments in north-eastern Sakhalin has become an acute problem (Figs. 48 and 49). In this regard, it is of scientific and practical interest to analyse the failure of barrier-type accumulative banks with a fixed upper part of the bank profile.

The problem was solved on the basis of detailed morpholithodynamic observations at all three sites and mathematical modelling of the dynamics of the barrier form (overflow), between the sea and the low marshy surface of the lagoon called Lake Mereya (Fig. 50).

Fig. 47 Remains of the former district centre of Kirovsky settlement. Kirovsky village, Kamchatka Territory (http://www.kamchadaly.ru/index.php/ kunena/sman_rodma/107-kirovskij-rybokombmat?start = 48)

Fig. 48 Temporary bank protection of approximately 800 m at the Chaivo Lagoon embankment (2003–2019)

In its present form, the embayment was probably formed during the last phase of Holocene coastal-marine sedimentation. During the first phases of mid-late Holocene sedimentation, lagoonal terraces up to 3.5 m high were formed during the ingression of marine waters into the Mereya River valley (Korotkiy 1997). The area of the estuarine lagoon was significantly larger during this period. This is evidenced by estuarine-lagoon sediments with a thickness of about 10 m 200 m west of the modern rooting base of the embankment and lagoonal sediments from a few metres

Fig. 49 Temporary pipeline protection at the Piltun Lagoon embankment (2016)

Fig. 50 Location of the Mereya Lagoon embankment

of uncovered thickness below the modern beach. The palaeodolin of the Mereya River can be traced in the bottom topography to depths of more than 30 m. The thickness of estuarine-lagoonal sediments filling the river valley increases along the traverse of the embankment from 7 m in the western part of the embankment to 17–18 m in the central part. In the area of the modern lagoon mouth, the thickness of these sediments decreases to 5 m and they are wedged out 180 m further to the east. During engineering surveys in the water area adjacent to the palaeodolin at depths of 15.5–17 m, 13–14 and 8 m, ancient shorelines were established, which indicates that the lagoon was substantially reorganised in the Early Holocene. The underwater coastal slope in the study area is composed of rather easily erodible rocks. The submerged watershed surfaces are expressed in the underwater relief by outcrops of bedrock mudstones of the Upper Cretaceous age of the Bykov Formation.

The bathymetric basis for the model is the navigational chart of Aniva Bay and measurements taken during our engineering and geomorphological surveys for shore protection purposes. The model is based on a navigational chart of Aniva Bay and measurements taken during our geomorphological engineering surveys for coastal protection purposes.

Wave parameters were determined by calculating the fields of frequency-directed spectra of sea swell based on wind fields recorded by a network of coastal meteorological stations over several decades. Data from the Sakhalin II Feasibility Study, adjusted for more recent studies (Project "Sakhalin II". Stage 2. Feasibility study. T. 5, book. 9, part 2. Environmental protection / M.: LLC NPF "Eco-center MTEA"—Sakhalin 2002), were also used. Here, the average deep water wave parameters (height $\bar{H}$ and period $\bar{T}$) corresponding to different wind speeds (W) are given for the main wave-prone channels (SW, SW and SE), and the total annual duration (tw) of wave events is also shown (Table 1).

The angles of wave approach Θ correspond to the angles between the wave direction and the north-south line. It is evident that the south-trending waves are significantly dominant in terms of duration and strength. The characteristics of extreme storms are reflected in Table 2.

The tide in Aniva Bay has a diurnal character. The tide in Aniva Bay is of a diurnal nature and its height was assumed to be 1 m. Storm surge heights can be up to 0.8 m with a recurrence period of 50–100 years, which was taken into account in the modelling of extreme storms (Table 2). Sand on the bottom and on the beach was considered as coarse-grained with a characteristic average size of 0.7 mm.

Modelling of wave impacts on the coastal profile was based on the CROSS-P model (Leontiev 2014a, b, 2015) using the parameters of extreme storms characterised in Table 2. The duration of storms was set equal to 12 h. The results obtained are presented in Fig. 51a. The initial profile and the profiles formed at the end of the storms are shown. The initial profile was taken for a shore section with a maximum elevation of 3.5 m.

As can be seen, with each storm the body of the embankment is eroded and the washed away material is carried to the underwater slope. Storms with a recurrence period of 1 year, 10 and 100 years cause retreat of the bank edge by 7, 15 and 22 metres respectively.

Table 1 Average wave parameters (height $\bar{H}$ and period $\bar{T}$) in deep water and their annual duration (t_w) for main directions

Rhumb	W, m/s	$\bar{H}$, m	$\bar{T}$, s	tw, hour
SW Θ = – 45°	5	0.3	2.7	543
	8	0.7	3.8	210
	11	1.1	4.8	64.2
	16	1.8	6.0	18.2
	20	2.5	6.9	5.1
	30	3.9	8.3	0.7
S Θ = 0°	5	0.3	2.7	596
	8	0.7	4.1	251
	11	1.2	5.2	48.9
	16	2.2	6.7	8.8
	20	2.9	7.6	1.5
	22	3.3	8.0	1.1
	26	3.9	8.7	0.7
SE Θ = 45°	5	0.3	2.7	280
	8	0.7	3.9	105
	11	1.2	5.0	34.3
	16	2.0	6.4	18.2
	20	2.7	7.2	4.4
	22	3.0	7.5	1.5

Table 2 Parameters of extreme storms of different frequency of occurrence

Parameter	Repeatability period, years		
	1	10	100
$\bar{H}$, m	1.8	2.5	3.3
$\bar{T}$, s	6.3	7.1	7.9
Surge, m	0.4	0.6	0.8

It is known that during periods of prolonged low wave action (quicksand) bottom material moves predominantly to the shore and the eroded beach can be restored to a certain extent. In order to clarify the applicability of this rule to the coast under consideration, modelling of the impact of weak tidal waves on the coastal profile formed by a once-a-year storm was carried out (Fig. 51b). The tidal parameters are as follows: mean height 0.4 m, period 8 s, duration 7 days.

Figure 51b shows that the ripples do transport sediment from the submerged slope to the beach where it is deposited in the form of a beach berm. However, the upper part of the eroded slope obviously cannot be recovered because the ripples do not reach it (due to insufficiently high surge and splash). In the next storm, the coastal berm will be rapidly eroded and the upper part of the slope will again be

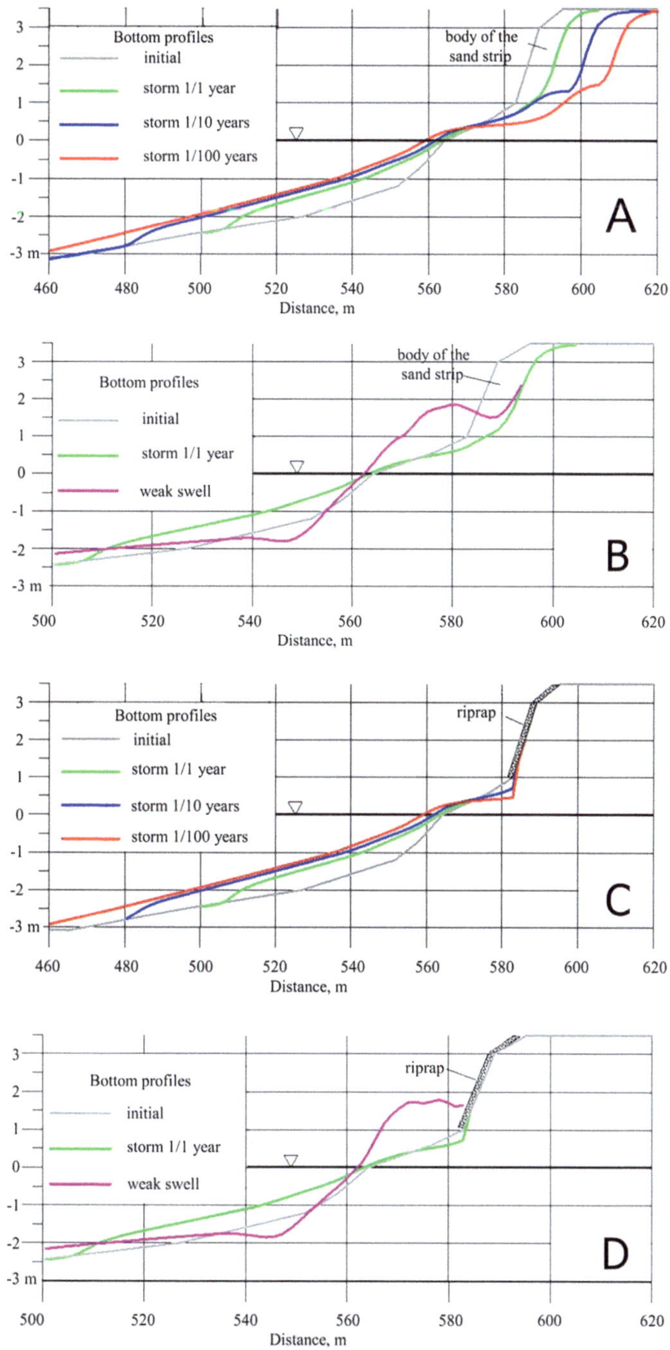

Fig. 51 Wave impacts on the shore profile of the Mereya Lagoon embankment: (**a**) deformations of the shore profile during storms of different strength; (**b**) impact of weak long-lasting quicksand on the storm profile; (**c**) storm deformations of the profile in the presence of rock pile; (**d**) impact of quicksand on the fixed slope

subject to erosion. The loss of material from the embankment body is thus irreversible.

Recently, rock revetment has been used to protect the seaward side of the embankment, which is considered in the model as a slope-type wave-absorbing protection. In this regard, modelling of storm impacts on the shore in the presence of slope bracing has been carried out. The calculations assume that the base of the anchorage is located at a beach elevation of +1 m and that material movement within the rock revetment is excluded. The results of the calculations are summarised in Figure 51c. Changes within the beach are found to be negligible. The most noticeable effect is a slight scour and bottom deepening at the base of the anchorage, which is about 0.2 m for the 1/1 year and 1/10 year storms and about 0.4 for the 1/100 year storm.

Some slumping of the anchorage material may occur as a result of scour. However, during periods of weak quicksand, the base of the anchorage will be backfilled with material coming ashore. This is evidenced by the modelling results shown in Figure 51c.

The sediment flux calculations were based on the LONT-2D model (Leontiev 2014a, b). Figure 52 shows the distribution of sediment fluxes generated by waves from different directions (SW, SE and SW) and the resulting flux, which is the

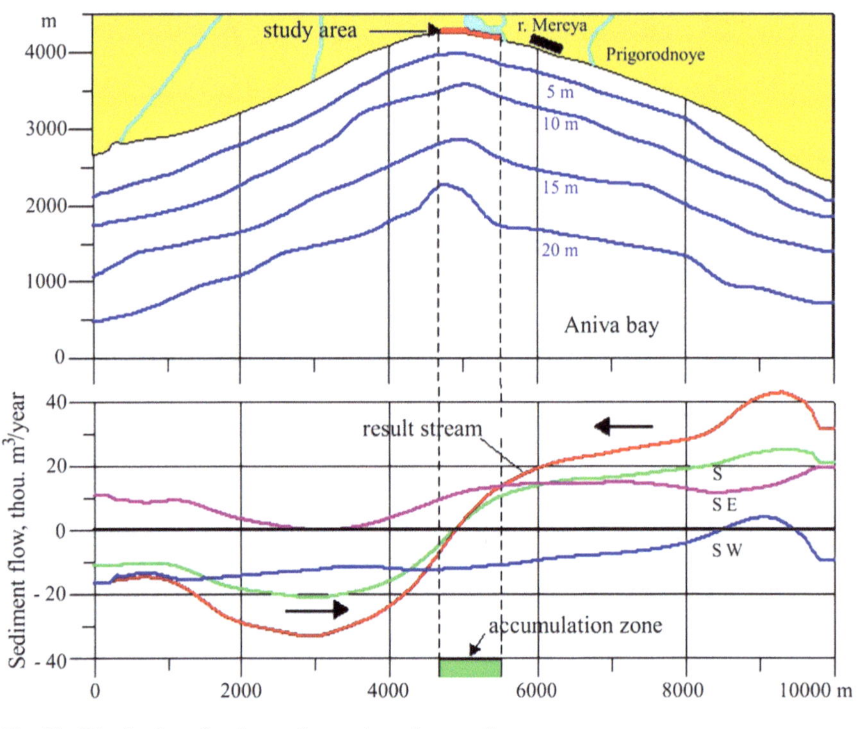

Fig. 52 Distribution of sediment fluxes along the coastline

algebraic sum of the individual fluxes. The eastward flow is considered negative and the westward flow positive.

SW waves push sediment eastwards, SW waves push sediment westwards, and the respective fluxes compensate for each other to some extent. Thus, southerly waves are responsible for the transport of sediment from the edges to the top of the bay. As the southerly waves dominate, the resultant transport has the character of two opposing currents converging at the top of the bay. Obviously, the considered area turns out to be in the convergence zone. As a result of unloading of the flows, accumulation of material should take place here. In particular, the western flow should accumulate material from the mouth of the Mereya River.

The values of individual fluxes vary within ±20 thousand m^3/year, and the amplitude of the resulting flux is close to 40 thousand m^3/year. The average flux gradient, which characterises the volume of accumulation in the study area, is of the order of 10 m^3/m −1 year −1, i.e. 10 m^3 per metre of bank length per year, which is not insignificant. However, these values should be treated with caution.

The point is that the calculations give the capacity of the currents, i.e. the volume that the waves could carry if there were enough sediment on the underwater slope and beach. However, in the conditions of the region in question, the sediment layer is not continuous: there are bedrock outcrops and boulder scatters over a large area of the seabed in the coastal zone. In other words, there is a deficit of material and the actual fluxes may be considerably less than those calculated.

Nevertheless, the current sediment fluxes contribute to the maintenance of the existing embankment. At the same time, there is no bank extension as a result of accumulation. Accumulation is therefore neutralised by processes of the opposite direction. One of such processes may be an increase in relative sea level as a result of tectonic land sinking and global ocean level rise.

Verification of the obtained information was carried out on the basis of trend analysis of the coastal zone. The reformation of the underwater relief is established on the basis of comparison of the 2004 and 2015 measurements (Fig. 53).

The data in dwg format were digitised in DNG system and translated into GIS. From the digitised data, points were triangulated and digital uniform grids were constructed at 1/70 arc second discreteness. A small amount of surface smoothing was then performed. The next step was to subtract the surfaces and create the resulting difference map. QGIS 2.6.2 software was used to construct the maps and all calculations were performed in ESPG:3395. The resulting residual relief surface map was saved as an image.

According to the data obtained, the sediment thickness increases over 77% of the submarine coastal slope area. Accordingly, the scour area is 23%. The input to the sediment budget over the time interval considered is 315,375 m^3, the volume of scour is 42,940 m^3 and the resulting input is +272,435 m^3 over 11 years, or 24,767 m^3/year. This value is fully consistent with the calculated sediment load capacity, as the equipment unloading berth is a barrier to the main sediment flow in an easterly direction.

The maximum surface depression of the submarine slope in the south-eastern part is approximately 2 metres. This is probably due to dredging in 2005–2006 to

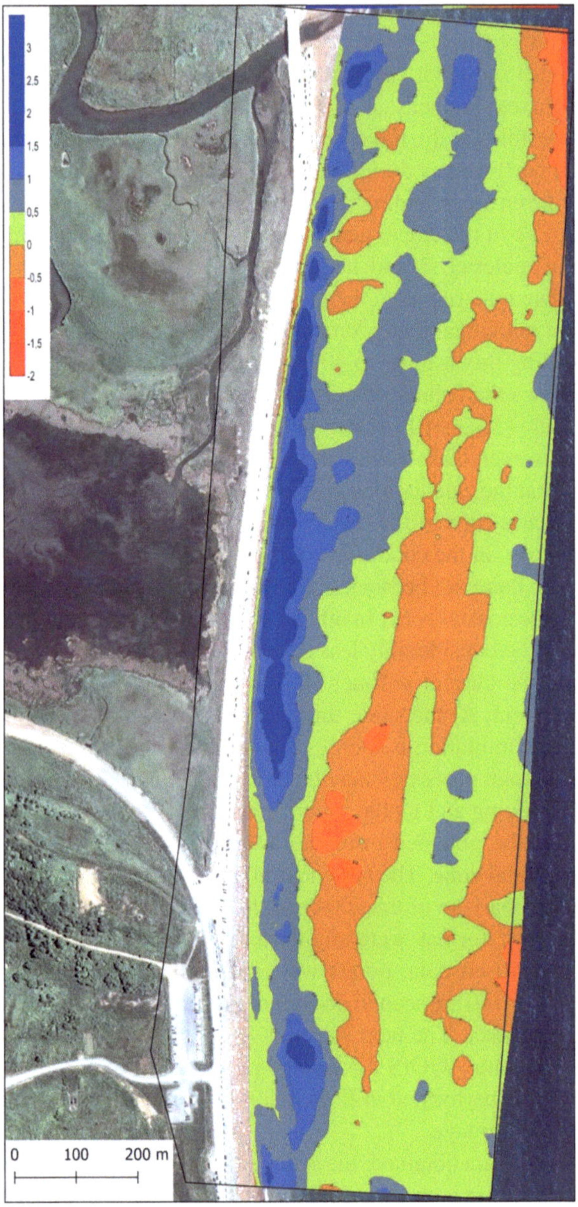

Fig. 53 Map of bank slope deformations based on comparison of 2004 and 2015 measurements

ensure navigational depths at the liquefied natural gas (LNG) jetty. The bulk of sedimentation is observed in the nearshore zone up to the 3 metre isobath. The minimum thickness of the active layer here is 2.5 metres.

Finally, a few words on the influence of changes in relative sea level on the dynamics of the embayment. Unfortunately, there are no reliable data on the rate of sea level change in the area under consideration. For a rough estimate, a value of w = 3 mm/year can probably be used, which is slightly higher than the average w value for the World Ocean. Further, we will use Bruun's rule to estimate the corresponding coastal retreat:

$$\frac{\partial x_0}{\partial t} = -\frac{w}{\overline{\beta}}, \quad \overline{\beta} = \frac{z_c + h_*}{l_X} \tag{1}$$

where $\partial x_0/\partial t$ —coastal drift rate, $\overline{\beta}$ —average bottom slope of the active part of the profile length l_X, located between the depth of closure h_* and the elevation of the beach z_c. The closure depth corresponds to twice the significant wave height observed for no more than 12 h per year. Judging from the available wave data, the desired wave height is close to 4 m, and therefore, $h_* = 8$ m. The coastal elevation can be taken as the average of the embankments $z_c = 3$ m. According to the measurements, $l_X = 550$ m.

As a result, estimation by the above formula gives the rate of coastal retreat $\partial x_0/\partial t$ = – 0,15 m/year. Obviously, the shore will retreat by only 1.5 m in a decade. Even if the value of w is doubled, the changes will be insignificant. The effect of relative level rise in this case will be noticeable only on the scale of hundreds of years. Thus, the factor of level rise hardly has a significant influence on the dynamics of the embankment.

The results of the modelling suggest that the wave-absorbing anchorage of the seaward slope of the embankment stabilises the shore profile parameters and does not alter the parameters of longshore sediment transport, as it is located above the zone of most active transport. The anchorage will therefore be sufficiently durable and the embankment body will be stable. At the present rate of relative sea level rise, this process has no discernible effect on the dynamics of the coastline under study and may only occur on a scale of centuries.

Calculations of sediment fluxes based on the lont-2d model, carried out in areas for which morpholithodynamic measurements are possible, are highly reliable. For example, the calculated sediment flow capacity in the area of the Mereya Lagoon overflow is about 20,000 m^3/year. Following the construction of the LNG jetty in 2005, an accumulative form of inlet corner fill with a volume of about 22,000 m^3 was formed on the western side of the LNG jetty within 9 months. Comparison of the 2004 and 2015 measurements showed that the sediment budget input from the western side of the jetty over the time interval considered is 315,375 m^3, the scour volume is 42,940 m^3, the resulting input is +272,435 m^3 over 11 years, or 24,767 m^3/year.

6 Conclusion

Thus, as a result of the conducted studies of seashore destruction it is established that:

1. Destruction of subvertical shore ledges in rocks of weak and medium stability under conditions of wave action occurs with velocities approximately one order of magnitude greater than the destruction velocities of subhorizontal surfaces of benches composed of similar rocks.
2. The measured retreat velocities of the coastal scarp without direct wave action approximately correspond to the fracture velocities of rocks of similar stability class on the benches. In both cases there are no shock wave loads with abrading effects of beach sediments. And if from Bench the weathering products are removed by sea water, in the case of cliff it occurs due to slope, mainly gravitational processes.
3. The main regularities of the spatial structure of the destruction of the banks of the Neopleistocene-Holocene weakly consolidated and loose sediments determine the rhythmicity phenomena in the manifestation of scouring processes. The rhythms of scouring with wavelengths of 1.3; 2.0; 4.0; 5.6; 6.5; 11.6; 25.6; 64 km have been established.
4. The analysis of morpholithodynamics of complex and "classical" sandy embayments separating lagoons from the open sea has shown that complex barrier forms are characterised by a higher variability of deformations of the marine edge of Holocene sections than fully Holocene embayments. Such barrier forms are also characterised by a greater variety of morphometric parameters of the upper part of the coastal profile. The lowest variability of morpholithodynamic parameters is characteristic of Holocene marine terraces attached to ancient coastal escarpments. These areas, located between lagoonal overflows, are least susceptible to scour.
5. The high-order rhythmicity of scouring processes, in our case ≈30–60 km, on the levelled shores of north-eastern Sakhalin, as well as on the shores of north-western Sakhalin, is associated with stable zones of divergence of resultant and seasonal longshore sediment fluxes with reduced beach sediment volume. Lower-order scour rhythms are, in our opinion, related, on the one hand, to processes of self-organisation of the coastal relief during seasonal changes in sediment transport directions, and, on the other hand, to peculiarities of the localisation and structure of feeding areas of lithodynamic systems. The 1.3 km scour rhythmicity corresponds to the dimensionality of the large megafeston structures of the beach.

References

Stephenson, W. J. Rock coasts. / W. J. Stephenson, M. E. Dickson, A. S. Trenhaile // Treatise on Geomorphology, Volume 10, Coastal Geomorphology. – Academic Press: San Diego, CA, 2013. – P. 289–307.

Sunamura, T. Rocky coast processes: with special reference to the recession of soft rock cliffs / T. Sunamura, // Proceedings of the Japan Academy. – 2015. – Series B, 91(9). – P. 481–500.

Trenhaile, A. S. Shore platform erosion and evolution: Implications for cosmogenic nuclide analysis / A. S. Trenhaile //Marine Geology. – 2016a. – T. 403. – P. 80–92.

Cruslock, Eva M. Geomorphologic equifinality: A comparison between shore platforms in Höga Kusten and Fårö, Sweden and the Vale of Glamorgan, South Wales, UK / Eva M. Cruslock, et al. // Geomorphology. – 114. – 1–2(2010). – P. 78–86.

Trenhaile, A. S. Hard-Rock Coastal Modelling: Past Practice and Future Prospects in a Changing World / A. S. Trenhaile //Journal of Marine Science and Engineering. – 2019. – T.7. – №. 2. – P. 34.

Emery, K. O. Sea cliffs: their processes, profiles, and classification / K. O. Emery, G. G. Kuhn // Geological Society of America Bulletin. – 1982. – T. 93. – №. 7. – P. 644–654.

Sunamura, T. Geomorphology of Rocky Coasts [Text] / T. Sunamura New York, Wiley, 1992 / – 302 p.

Griggs, G. Sea cliff erosion / G. Griggs, L. Savoy // Living with the California Coast. – Durham: Duke University Press, 1985. – P. 27–34.

Polunin, G. V. Map of lithological complexes and manifestations of exogenous processes of Sakhalin: scale 1:500000 / G. V. Polunin, V. A. Buzlaev. - M.: GUGK, 1984.

Polunin, G. V. Map of lithological complexes and exogenous processes of the Sakhalin land and shelf on a scale of 1:500000 / G. V. Polunin. - Khabarovsk, Yuzhno-Sakhalinsk, 1992.

Golozubov, V. V. (2012). Deformations of Late Cretaceous and Cenozoic complexes of the West Sakhalin terrane / V.V. Golozubov, S.A. Kasatkin, V.M. Grannik, A.E. Nechayuk // Geotectonics. - (5). - With. 22.

Trenhaile, A. S. Shore platform downwearing in eastern Canada A 9 – 14 year micro-erosion meter record / A. S. Trenhaile, N. J. Porter // Geomorphology. – 2016b. – T. 311. – P. 90–102.

Trenhaile, A. S. Shore platform erosion and evolution: Implications for cosmogenic nuclide analysis / A. S. Trenhaile //Marine Geology. – 2016c. – T. 403. – P. 80–92.

Porter, N. J. Patterns of surface downwearing on shore platforms in eastern Canada / N. J. Porter, A. S. Trenhaile, K. Prestanski, J. I. Kanyaya // Earth Surface Processes and Landforms. – 2010. – 35(15). – P. 1793–1810.

Stephenson, W. J. «Forty three years of micro-erosion meter monitoring of erosion rates on shore platforms at Kaikōura Peninsula, South Island, New Zealand. « / W. J. Stephenson, R. M. Kirk, M. A. Hemmingsen // Geomorphology. – 344 (2019). – P. 1–9

Stephenson, W. J. 2017. New insights on the relative contributions of coastal processes and tectonics to shore platform development following the Kaikōura earthquake / W. J. Stephenson, M. E. Dickson, P. H. Denys // Earth Surface Processes and Landforms. – 42(13). – P. 2214–2220

Afanas'ev, V. V. Processes of destruction of the seashore of the newly formed at Mw=6, 2 Nevel earthquake 2007 / V. V. Afanas'ev, L. M. Bogomolov // Trigger effects in geosystems: theses of reports of the Second All-Russian seminar-conference. - Moscow, IDG RAS, 18-21 June 2013 - P. 5-5.

Westoby, M. J. Structure-from-Motion'photogrammetry: A low-cost, effective tool for geoscience applications / M. J. Westoby, J. Brasington, N. F. Glasser, M. J. Hambrey //Geomorphology. – 2012. – T. 179. – C. 300–314.

Afanas'ev, V. V. Geomorphological aspects of the problem of site selection for the construction of liquefied natural gas plants on the coast of the Far Eastern seas / V. V. Afanas'ev // Proceedings: XXXVI Plenum of the Geomorphological Commission of the Russian Academy of Sciences All-Russian scientific and practical conference with international participation "Geomorphology - science of the XXI century". - (Barnaul, 24-28 September 2018). - Barnaul. Publishing house of Altai State University, 2016. - P. 26-31

Okamura S. Cenozoic magmatism of the North-Eastern Eurasian margin: the role of lithosphere versus asthenosphere / S. Okamura, R. J. Arculus, Y. A. Martynov // Journal of Petrology. – 2004. – T. 45. – №. 2. – P. 221–253

Map of volcano-tectonic structures of the coastal-continental part of the Far East of the USSR. 1:1500000 / Ch. ed. N. A. Shilo, Yu. A. Kosygin. 1982.

Afanas'ev, V. V. Problems of coastal use of subarctic seas (on the example of Sakhalin Island) / V. V. Afanas'ev // Modern problems of utilising the potential of marine areas and coastal zones. - 2015. - C. 118-130.

User Manual Agisoft PhotoScan Professional Edition, version 1.2 [Text] / 2015. – 116 p.
Bemis, S. P. Ground-based and UAV-based photogrammetry: A multi-scale, high-resolution mapping tool for structural geology and Paleoseismology / S. P. Bemis, S. Micklethwaite, D. Turner, M. R. James, S. Akciz, S. T. Thiele, H. A. Bangash // J. of Struct. Geol. – 69(2014). – P. 163–178
Afanas'ev, V. V. Coastal erosion of north-western Sakhalin / V. V. Afanas'ev // Coastal zone of the Far Eastern seas. - L.: Izd-vo GO USSR, 1991. - P. 98-104
Dolan, R. Storms and shoreline configuration / R. Dolan, B. Hayden // J. Sediment. Petrol. – 1981. – V51. – No 3. – P. 737–744.
Solov'ev, I. A. Channel process and waterways of the Amur estuary [Text] / I. A. Solov'ev. - Vladivostok, 1974. - 290 p.
Popov, B. A. Correlation of the dynamics of the shores of the Arctic and southern seas / B. A. Popov // Palaeogeography and Pleistocene sediments of the southern seas of the USSR. - Moscow: Nauka, 1977. - P. 171-179.
Zenkovich, V. P. About some theoretical provisions of coastal science / V. P. Zenkovich, R. Y. Knaps // Theoretical problems of sea coast development. - Moscow: Nauka, 1989. - P. 78-82.
Knaps, R. Ya. To determination of the characteristics of the energy state of the coastal slope / R. Ya. Knaps // Studies of the sea coast relief dynamics. - Moscow: Nauka, 1979. - P. 70-80.
Piskareva, M. A. Comparability of some energetic methods of estimation of longshore sediment transport / M. A. Piskareva, I. A. Tretyakova // M., 1986. - Dep. Herald of Moscow State University. - Series 5, Geography. 5460. - B86. - 10 p.
Selivanov, A.O. Experience of statistical study of the relationship between the character of the wave-tidal field and features of the coastal zone cross-sectional profile / A.O. Selivanov, L.V. Anzimirov // Oceanology. - 1985. - VOL. XXVI. - Vol. 5. - P. 806-812
Yurkevich, M. G. Investigation of the lithodynamic regime of the coastal zone of the shelf of the north-east Arctic / M. G. Yurkevich // Natural bases of coastal protection. - Moscow: Nauka, 1987. - P. 138-146.
Popov, B. A. Some features of the dynamics of the Arctic shores of Asia / B. A. Popov, V. A. Sovershaev // Marine shores. Questions of geography. - Moscow: Mysl, 1982. - P. 105-115.
Popov, B. A. Principles of selection of initial data for calculation of wave energy load / B. A. Popov, V. A. Sovershaev // Coastal zone of the sea. - Moscow: Nauka, 1981. - P. 47-52.
Tikhonovich, N. N Geomorphological sketch of the Russian Sakhalin. Tr. Geol. com. Nov. Ser. Vp 120 / N. N. Tikhonovich, I. I. Polevoy. - Pg.: Tip. M. M. Stasyulevich, 1915. - 77 p.
Vladimirov, A. T. Morphology and evolution of the lagoonal shore of Sakhalin Island / A. T. Vladimirov // Proc. of the Institute of Oceanology of the USSR Academy of Sciences. Institute of Oceanology, USSR Academy of Sciences, vol. 46. - M.: ANS SSR, 1961. - P. 145-171.
Buzlaev, V. A. New data on the Sakhalin coastal processing rates / V. A. Buzlaev // Geographical studies of the Far East seas shelf. - Vladivostok: Izd. v. Dalnevost. un-ta, 1993. - P. 78-85.
Remenets, A. I. Evolution of the lagoonal coast of Piltun Bay in the Holocene / A. I. Remenets, I. N. Klimov // Development of the natural environment in the Pleistocene. - Vladivostok: Far Eastern Scientific Centre of the Academy of Sciences of the USSR, 1981. – P. 123-125.
Volodarskiy, A. N. Dynamics of lagoon shores of Sakhalin in the areas of active shelf development / A. N. Volodarskiy, Y. A. Mikishin, M. V. Cherepanova // Complex problems of study, development and protection of mineral resources of the ocean. - Vladivostok: Far Eastern Scientific Centre of the Academy of Sciences of the USSR, 1983. - P. 24-25.
Afanas'ev, V. V. Evolution of the Far Eastern seas coast in the Holocene / V. V. Afanas'ev // Coastal evolution in the conditions of ocean level rise. - M.: Izd-vo MSU, 1992. - P. 166-174.
Mikishin, Yu. A. Stratigraphy of sediments and palaeogeography of the north-eastern Sakhalin coast in the Holocene / Yu. A. Mikishin, I. G. Gvozdeva // Nauchnoe obozrenie. - 2005. - № 3. - P. 4-15.
Safyanov, G. A. Geomorphology of sea shores [Text] / G. A. Safyanov. - Moscow: Izd-vo MSU, 1995. - 400 p.

Zenkovich, V. P. From the foreign experience of marine shore protection / V. P. Zenkovich // Natural bases of shore protection. - M.: Nauka, 1987. - P. 149-153.
Korotkiy, A. M. Marine terraces and Quaternary history of the Sakhalin shelf [Text] / A. M. Korotkiy, V. S. Pushkar, T. A. Grebennikova. - Vladivostok: Dalnauka, 1997. - 195 p.
Sakhalin II Project. Phase 2 Feasibility Study. T. 5, Book 9, part 2. Environmental Protection / M.: OOO NPF "Eco-Centre MTEA" - 2002.
Leontiev, I. O. Modelling of storm deformations of a sandy shore (by the example of the eastern part of the Gulf of Finland) / I. O. O. Leontiev, D. V. Ryabchuk, A. Yu. Sergeev // Oceanology. - 2015. - T. 55, № 1. - P. 147-156. - (https://doi.org/10.7868/s0030157414060069)
Leontiev, I. O. On the calculation of longshore sediment transport / I. O. O. Leontiev // Oceanology. - 2014a. - T. 54, № 2. - P. 226-232. - (https://doi.org/10.7868/s0030157414020130)
Leontiev, I. O. Morphodynamic processes in the coastal zone of the sea / I. O. Leontiev. - LAP LAMBERT Academic Publishing. Saarbrücken, Deutschland (Germany), 2014b. - 251 p.

Beach Formation and Distribution of Its Wave-Absorbing Properties along the Shore

V. Afanas'ev Viktor and E. I. Ignatov

Abstract This study investigates beach morpholithodynamics within the context of superimposed wave and tidal influences. Findings from beach profile surveys are analyzed, with a specific focus on beaches exhibiting an uneven alongshore distribution of sediment due to the development of megafestoons (large-scale sand waves) and their associated systems.

The paper presents findings from morphometric monitoring (2005–2022) of a stable megafestoon system that has persisted for at least 65 years. Key morphometric and morphodynamic parameters for this system are established, including deformation volumes and changes in its spatial configuration over specific time intervals.

While the precise causes and mechanisms for this alongshore heterogeneity remain unclear, several observations provide insight: the spatial instability of system components, sediment deformations extending to the depth of closure, and the alongshore migration of accumulative features in the direction of longshore sediment transport. These factors suggest that the system represents a self-organized response to spatiotemporal variations in wave approach angle. In contrast, it is likely that stable megafestoon systems on open ocean coasts form through different mechanisms.

Keywords Shoreline contour · Cuspate bumps · Circulation cells · Selforganization of bed relief · Morphometry of the beach · Megacusps · Sandwaves

V. Afanas'ev Viktor (✉)
Sakhalin State University, Yuzhno-Sakhalinsk, Russia

E. I. Ignatov
Faculty of Geography, Moscow State University, Moscow, Russia

T. Chaplina (ed.), *Coastal Geosystems of Cold Seas Under Conditions of Climatic and Anthropogenic Changes*, Earth and Environmental Sciences Library, https://doi.org/10.1007/978-3-032-12247-6_4

1 Introduction

The peculiarities of hydrodynamic conditions, the nature of sedimentation and relief formation have led to the emergence and development of the concept of vertical zonality of the coastal zone. The analytical review of the existing ideas made by Y.S. Dolotov (1989), as well as a large number of generalisations devoted to various aspects of the problem and based on the results of numerous field studies, laboratory experiments and numerical modelling, allow to explain sufficiently the peculiarities of morphodynamics of the upper part of the coastal slope (Belberov 1984; Shuisky 1985; Leontiev 1989; Kosyan 1991; Aibulatov 1990; Horn 2002).

At the same time, it should be noted that the prediction of coastal slope deformations with known wave-tidal field parameters and sediment balance characteristics is still quite uncertain, so it is of great importance to analyse trends in the development of the submarine coastal slope and beach based on the results of in-situ studies of short-term, seasonal and multi-year changes (Fig. 1).

The most commonly used definition of 'beach' is as follows: a beach is an elementary accumulative form formed in the surf zone. In addition to the fact that a beach is a highly dynamic landform of the coastal zone, developed under the action of flat turbulent water flow with a regular change of direction, the sediments that compose it are part of the entire morpholithodynamic coastal system. And in areas where there is longshore sediment transport, the beach zone is one of the most energetic elements of sediment flow (Leontiev 1961; Díez, 2017). The amount of transport in the beach zone is difficult to quantify (Osborne 1999; Hughes 1997). However, sediment transport in the splash zone is known to be significantly higher than in the inner surf zone (Beach 1991; Elfrink 2002; Erikson 2004).

According to existing concepts, a full profile beach has, in general, 5 main morpholithodynamic zones (Dolotov 1982; Leontiev et al. 1974). An underwater stage of the beach is formed at the site of wave breaking and the origin of the sediment flow (Fig. 2). The step transitions into a narrow zone with a steeply sloping bottom, which corresponds to the area of highest storm-driven flow velocity—this is known as the beach scarp. Further up the profile, the beach scarp gives way to the on-drift beach step, or the berm. The berm then slopes upward, forming the backshore.

The berm itself develops where the final breaking wave collides with the backward wash of water from the previous wave. On its seaward slope, a zone of maximum sediment movement is observed. When sediment is scarce, this area appears as a concave shape or even a small scour ledge in the profile. Conversely, when sediment is abundant, the profile becomes convex.

Finally, the splash zone delivers sediment to the upper, dry part of the beach (the subaerial zone). From here, the wind can then transport this material further landward, supplying the dunes at the back of the beach.

Fig. 1 Dynamic condition of the beach at 106 km of the Yuzhno-Sakhalinsk—Okha—Moskalvo Road: (**a**) January 2000; (**b**) July 2000; (**c**) June 2001; (**d**) August 2004; (**e**) March 2009; (**f**) August 2014

2 Structure and Dynamics of the Upper Part of the Coastal Slope

Research on beach hydrodynamics and morphology carried out in the 1970s and 1980s led to the development of a morphodynamic model of the beach and the identification of six different morphodynamic states of the beach (Wright 1984). Two main profile types, reflective and scattering, and 4 intermediate profile types with reflective and scattering elements were identified.

Provided that the morphology is in equilibrium with the hydrodynamic forcing, dissipative, intermediate and reflecting beaches can be distinguished on the basis of the coefficient $\Omega = Hb/(wsT)$. Where Hb is the wave height, T is the wave period, and ws is the sediment deposition rate. It was clarified that dissipative beaches have

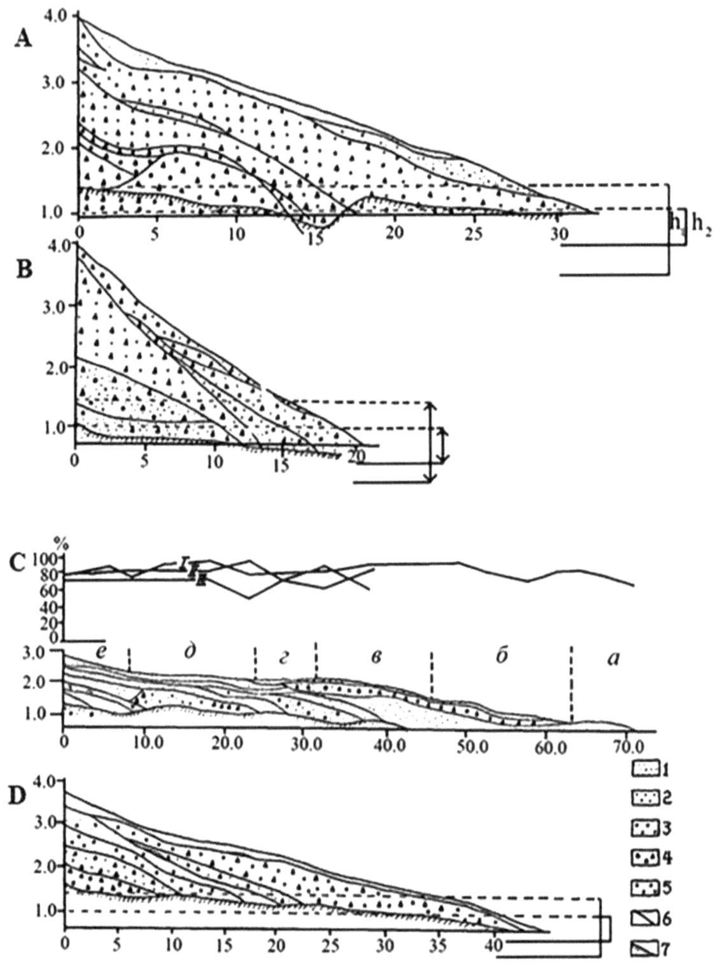

Fig. 2 Beach structure (based on the results of analyses of beach sections): A—area of coastal scarp erosion with velocities up to 0.2 m, B—area of coastal scarp erosion with velocities of 1.0-2.0 m, C—accumulative area, D—area with overgrown coastal scarp. 1—fine-grained sand, 2—medium-grained sand, 3—coarse-grained sand, 4—gravel, 5—pebbles, 6—facies boundaries, 7—groundwater table; h1—sisigian interval of tidal oscillations, h2—quadrature interval of tidal oscillations. I—profile distribution of surface sediments, II—profile distribution of modal fraction (0.1–0.25) of aeolian sediments, III—profile distribution of modal fraction (0.25–0.5) of storm cycle sediments. Morphodynamic zones of the beach: a—nearshore part, b—tidal zone of the sea slope, c—sea slope, d—top of the coastal berm, e—aeolian accumulation zone, f—foot of the coastal scarp or accumulative terrace

$\Omega > 6$, reflective beaches form when $\Omega < 1.5$, while intermediate beach states occur in the range of these values (Short 1999). The above paradigm of morphodynamic beach classification has been further confirmed and extended by studying shores with different wave intensity and tidal values (Castelle 2007; Masselink 1993; Short 1993).

The morpholithodynamic content of the state-reflecting beach transition scheme through all four intermediate types to dissipative is presented in T. Aagaard et al. (2013), but it should be noted that the splash zone is not included in this beach state model.

It is noted that despite the emerging work in this area, there is still insufficient information to add sediment transport in the splash zone to a conceptual model of beach states (Hughes 2014).

The splash zone is the most terrestrial part of the system and supplies sediment to the subaerial beach, which is then available for aeolian transport to accumulative terrace and dune systems. Our studies have shown that aeolian transport is possible at velocities greater than 15 m/s directly from the nearshore wet part of the beach, and that aeolian sediment spectra are found almost everywhere in coastal berms and beach moulds (Figs. 2 and 3).

Recent observations on the shores of the Netherlands have shown that there is no significant correlation between the variability of annual wind conditions) and annual changes in dune volume. Conversely, a significant temporal correlation was found between variability in annual beach condition and changes in dune volume (De Vries 2012, 2014). This suggests that the use of traditional aeolian morpholithodynamic models developed for desert conditions leads to an overestimation of the

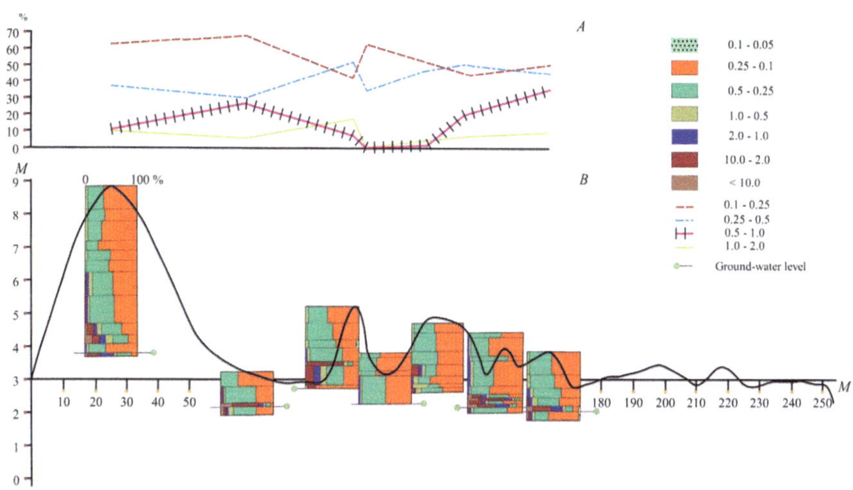

Fig. 3 Section of the system of coastal levees and foredunes on the coast of Aniva Bay. (**a**) distribution of granulometric dimensions along generalised shape transects, B—aeolian granulometric spectra in the sediments of the coastal ramparts and avandunes

importance of variability in wind conditions for values of sand transport across the beach towards avandunes and coastal terraces.

Based on previously established direct relationships between the modal (morpholithodynamic) state of the beach, the type and parameters of avandunes and the morphometry of the subaerial segment of the beach, four main morphotypes of 'dry' beach were identified (Fig. 4) (Dean 1994; Díez, 2016, 2017; Hesp 1986; Short 1979,1982).

Thus, it is the subaerial segment of the beach that, along with its function as the main wave-absorbing element of the coastal zone, is the source of silty psammite material to increase the contrast and size of coastal terraces and dunes.

This study analyzed the structure and dynamics of beach zones across multiple locations, including Ugolnaya Bay (Bering Sea), the Sakhalin coast of the Amur Liman and Nevelsky Strait, sections of the Tatar Strait and the Sea of Okhotsk, and Iturup Island.

Using repeated leveling profiles, we studied changes in beach morphometrics—such as shape and slope—across different timescales, from single storm events to seasonal and multi-year cycles. Beach width was also measured over long stretches using aerial photography interpretation. All analyses accounted for factors like the tidal phase during measurement and the preceding morphodynamic conditions.

The study of beach sediment structure and composition, conducted using the beach transect method, enabled precise identification of the lower boundary of the

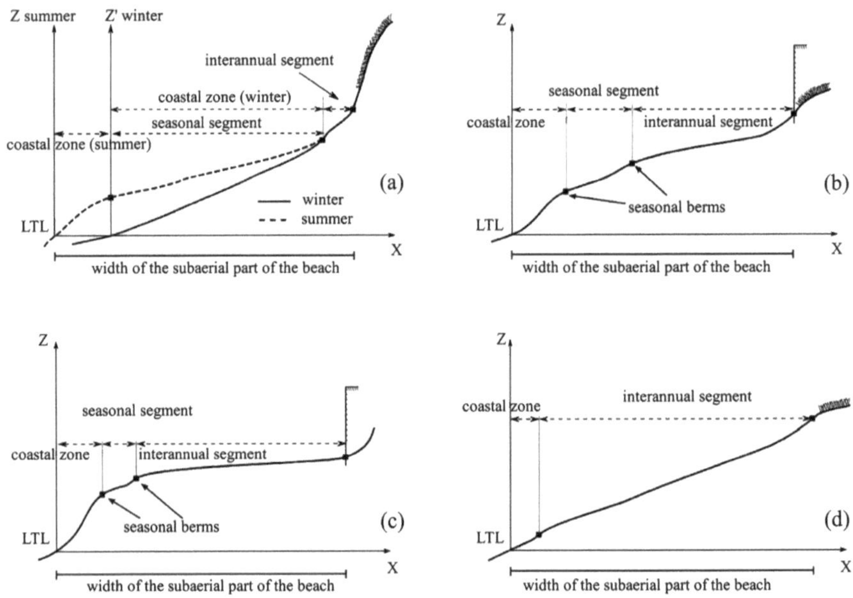

Fig. 4 Four morphotypes of subaerial profiles (from Díez et al. 2017). Type (**a**) refers to dissipative beaches subject to seasonality in wave climate, type (**b**) to intermediate, type (**c**) to reflective, type (**d**) to ultradissipative

seasonal active layer—a feature that is typically undetectable through short-term observational methods such as profile leveling. This approach also facilitated assessment of the sediment composition throughout the entire beach body, as well as within specific facies formed under distinct environmental conditions.

Furthermore, the use of beach transects with reference horizons allows reconstruction of earlier beach morphometric parameters, providing insight into conditions prior to the current observations. A significant volume of data was acquired during an engineering geomorphological study of approximately 30 km of beach, carried out as part of coastal reinforcement planning. This included hundreds of boreholes, which helped estimate the thickness of the active layer, calculate the total volume of beach sediments, and characterize the underlying surface.

One of the most thoroughly studied areas is the coast of Terpeniya Bay near Vzmorye, which has been monitored through drilling and morphometric observations for over 12 years. It is in this location that the stable megafestoons discussed in subsequent sections of the chapter have been documented.

A large amount of research was carried out on the shores of the Amur Liman and Nevelsky Strait (Afanas'ev et al. 2006). Seasonal features of beach morpholithodynamics determined by the character of the wind-wave regime were determined, which are expressed as follows. During the summer period, sediment accumulation occurs north of the mouth of the Pogibi River and south of the causeway. In areas adjacent to Cape Pogibi, beach sediments are eroded during this period. The excitation by autumnal, predominantly northerly winds forms a zone of accumulation to the north of the cape and to the south promotes the redistribution of sediment to a higher hypsometric level.

Analyses of the morphodynamic effect of specific hydrodynamic conditions in the area of Cape Pogibi. Analysis of the morphodynamic effect of specific hydrodynamic conditions in the vicinity of Cape Pogibi showed that seasonal trends in beach development are not determined by their simple sum (Afanas'ev 1996). A significant influence on beach formation processes is exerted by regular changes in the basin level as a result of tides, which have an irregular semidiurnal pattern in the area of Cape Pogibi. Pogibi is of irregular semi-diurnal character. The calculated mean annual level is 0.72 m relative to the lowest theoretical level (LTL). The height of the sisigian tide reaches 210 cm. and the quadrature tide reaches 90 cm to LTL. The highest tides are recorded in August-September. It is thought that the most general trend in the development of the beach profile under low tidal sea conditions is an increase in its steepness due to the redistribution of sediment from low to upper hypsometric levels (Komar 1975).

We have confirmed that, in the absence of inputs from adjacent sites, there is an increase in steepness at virtually all stations in the vicinity of Cape Pogibi. The steepness of the profile increases at almost all sites in the vicinity of Pogibi. However, the shape of beaches with a high content of gravel and pebble material does not change, as well as the shape of low thickness 0.10–0.15 m beaches, the shape of which is determined by easily erodible silt and clay at the base.

In coastal systems with active sediment transport and varying tidal ranges, waves approaching from a consistent direction can produce different beach formation outcomes depending on tidal phase.

During syzygy tides (spring tides), sediment transported from the north results in accumulation both at the river mouth and in the wave shadow area behind the headland. When sediment is supplied from the south, accumulation occurs not only in these sheltered zones but also within the updrift (inlet) corner of the coastal dike.

Under quadrature tides (neap tides), this pattern reverses. Southward sediment transport erodes areas near the headland while depositing material north of the Pogibi River mouth. During northerly swell conditions under quadrature tides, the overall pattern resembles that of syzygy tides, but the accretion zone extends slightly farther south.

The picture is completed by analysing the beach transects (Fig. 5). A two-layered beach structure is observed in Section 1, where scouring occurs under almost all conditions. The beach sediments themselves are 10–30 cm thick, i.e. the entire thickness is formed during one tidal cycle. The profile is characterised by the highest steepness for the area—5.4. Repeated levelling has made it possible to establish that even after storms of insignificant strength, the underlying sediments are eroded (up to 20 m), with the thickness of the beach sediment itself remaining virtually unchanged. The distribution of fractions on this profile varies from unimodal in the upper part to bimodal in the lower part, due to the removal of the fine fraction by the return flow.

On the submarine slope adjacent to the cut-off, the distribution is unimodal with a significant predominance of medium and fine sand fractions. In the area adjacent to the coastal slope, the sediment is characterised by a bimodal distribution. Of

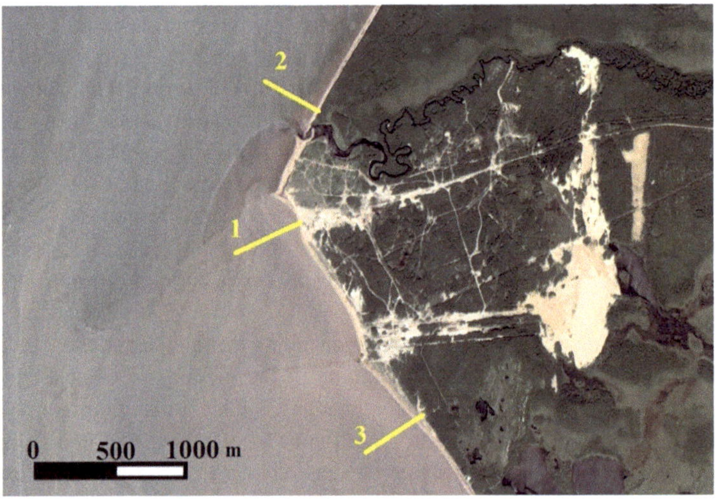

Fig. 5 Location of beach transects at Cape Pogibi (Nevelsky Strait. Pogibi (Nevelsky Strait). Photo from Bing public resource

particular note is that the gradients of change in the content of the 10–2.0 and 0.5–0.25 mm and 2.0–1.0 and 0.25–0.1 mm fractions are equal but opposite in sign at different elements of the beach profile.

Section 2 characterises the beach sediments, which are characterised by accumulation during quadrature level changes and scour during sisygian changes. The average gradient of the surface slope is 5° (from the cut-off to the beach berm and has almost identical parameters to the profile in section № 1. Beach sediments in the section are clearly differentiated into two layers. The lower layer consists of multigrained sands with a high content of gravel and pebbles. The thickness of the sediments reaches 80–100 cm. In the middle part of the profile, fragments of a half-destroyed coastal rampart are observed. The results of profile levelling in the season preceding these observations and the nature of the sediments indicate that this layer was formed during storms.

The upper layer is represented by sediments of finer granulometric size. The thickness of the layer is 20–50 cm. Compared with the sediments exposed in Section 1, this layer has a significantly lower content of 0.25–0.1 mm sediment and a higher content of gravel, with approximately the same composition in the 0.25–2.0 mm range.

Even more significant differences are noted when comparing the composition of sediments in profile sections with similar morphometric parameters. In this case, the gravel content is 40%, which is 2.5 times higher than in the sediments of Section 1, and the 2.0–1.0 mm fraction is 1.4 times higher. The content of 0.5–0.25 mm sediment is 1.8 times lower, 0.25–0.1 mm—5 times lower. The coarse sand content in both cases is approximately 12% each.

Thus, beach accretion is predominantly driven by coarse sediment fractions. Repeated sampling of this section showed that beach accretion is accompanied by an increase in sediment content of ≥1 mm and a decrease of ≤0.5 mm, with the original and accretionary beach containing approximately the same amount of sediment of 1.0–0.5 mm. At 500 m south of the dyke, section 3 reveals fairly homogeneous beach sediments, which increase in thickness from 0.3–0.4 m in the upper parts of the profile to 1.5 m in the lower parts. Dimensions 0.5–0.25 mm and 2.0–1.0 mm, as well as 1.0–0.5 mm and 2.0 mm form pairs of fractions, which at approximately equal gradients have the opposite direction of content change along the profile. At the same time, at low levels of the profile the content of sediments prevailing along the profile is lower than the average values, while at the upper levels it is higher. The beach located 6 km north of Pogibi Bay has the lowest gradients and the greatest width in the area of Nevelsky Strait. The beach transect ditch was preserved for 2 years, which is not typical of any other site. The beach sediments proper, stratified into two strata, are laid down on multigrained sands with traces of cryodislocations. The exposed thickness of sediments of the lower layer is up to 0.6 m, of the upper layer—up to 0.8 m.

The fractional composition of both strata is approximately the same. Comparing the sediments of the upper layer with those of the previously considered sites, one can note poor sorting of sediments and a decrease in the content of the main nannofossil fraction (0.5–0.25 mm). Equal gradient pairs with opposite sign make up the fractions: 0.5–0.25 and 2.0–1.0 mm and 10.0–2.0 and 1.0–0.5 mm.

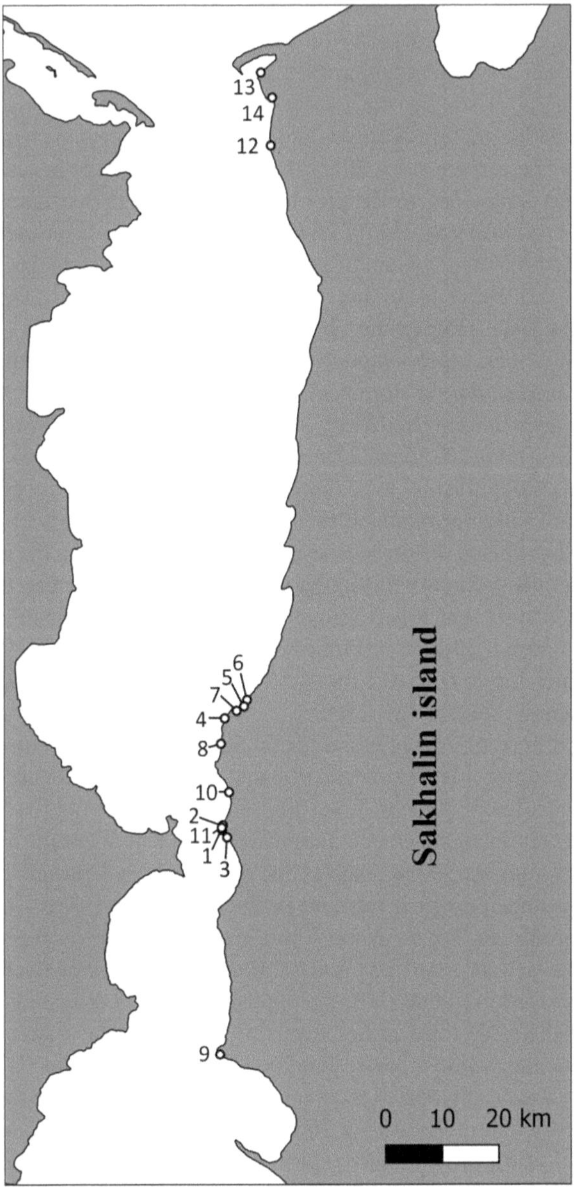

Fig. 6 Location of beach transects

Beach transects made during one sisigian phase of tidal fluctuations from Lakh to Rybnovsky provided information suitable for comparative characterisation of the beach zone of most of the shores of north-west Sakhalin (Fig. 6).

It was found that the main beach-forming fraction is medium-grained sand (0.5–0.25 mm), the content of which varies from 30 to 70% (Table 1, Fig. 7).

Table 1 Granulometric spectra of beach sediments by sections (whole thickness). A—layer formed by moderate waves with aeolian accumulations in the rear part; B—layer formed in storm conditions, with remnants of storm shafts; C—underlying sediments continuing under the coastal scarp

			Fraction size and content %						Coefficients				
№	Section	Layer	10,0–2,0	2,0–1,0	1,0–0,5	0,5–0,25	0,25–0,1	0,1–0,05	M1	M1	M3	o	к
1	I	A	0	5	33,1	53,8	8,1	0	0,3	0,41	0,64	0,68	1,12
2	II	A	28,3	13,7	14,5	38,7	4,8	0	0,33	0,7	2,4	0,37	1,61
3		Б	45,8	16,7	11,7	22,3	3,5	0	0,48	1,7	4,4	0,33	0,73
4	III	A	17,7	11,7	12,5	39,7	18,4	0	0,27	0,4	1,4	0,44	2,37
5	IV	A	14	15,6	20,8	41,3	8,3	0	0,31	0,5	1,3	0,49	1,6
6	V	A	0	1,9	4,6	65,3	28,2	0	0,24	0,32	0,42	0,76	1
7	VI	A	7,6	11	12,4	60,5	8,5	0	0,3	0,39	0,66	0,67	1,33
8	VII	A	20,5	20,1	13,3	36,9	9,2	0	0,31	0,6	1,7	0,43	1,47
9	VIII	A	2	8,2	25,8	63,7	0,3	0	0,33	0,44	0,68	0,7	1,16
10		Б	0,1	15,5	52,3	31,9	0,2	0	0,44	0,63	0,9	0,7	1
11		B	3,4	12	24,7	59,6	0,3	0	0,33	0,45	77	0,65	1,25
12	IX	A	3,4	5,9	17,4	65,6	7,7	0	0,29	0,36	0,49	0,76	1,08
13	X	A	19,9	21,5	25,6	30,6	2,4	0	0,4	0,77	1,7	0,48	1,15
14		Б	18,6	28,8	15,4	29,6	6,9	0,7	0,34	0,94	1,8	0,43	0,69
15	XI	A	8,9	14,9	21,8	43,7	10,6	0,1	0,3	0,46	0,98	0,55	1,38
16		Б	1,8	5,4	11,9	39,4	32,5	9	0,16	0,29	0,46	0,59	0,88
17	XII	A	7,5	30,1	20,3	38,4	4,2	0	0,36	0,68	1,4	0,51	1,09
18	XIII	A	7,6	18,8	22,5	37,1	13,8	0	0,3	0,5	1,1	0,52	1,32
19	XIV	A	17,2	13	9,7	26,1	34	0	0,2	0,36	1,4	0,38	2,15

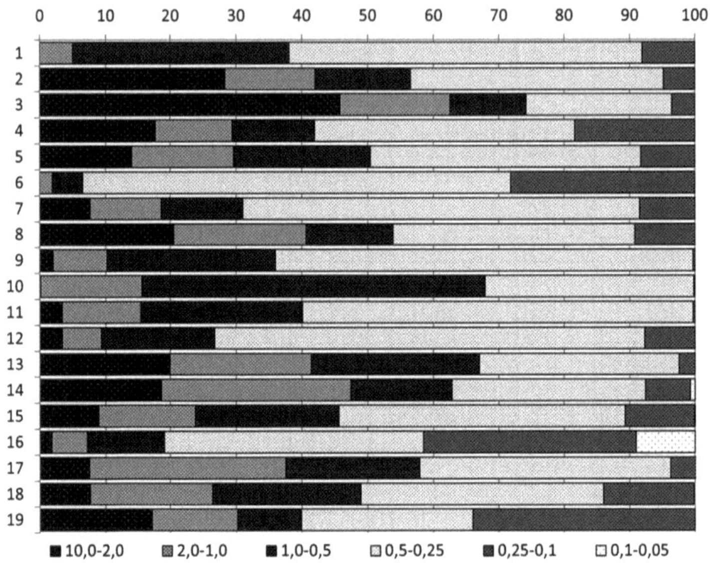

Fig. 7 Granulometric spectra of beach sediments in the north-west of Sakhalin Island (averaged over the entire section). Sakhalin (averaged over the entire section) visualisation of Table 1

The sediments in sections 10, 4, 11, 13 and 14 are characterised by an opposite sign of change in the distribution of sediment of the dominant (0.5–0.25 mm) and (2.0–1.0 mm) size fractions across the profile. In sections with a tendency to accumulate, there is a synchronised change in the content of sediment of the predominant fraction and the next largest size (1.0–0.5 mm). In the transit sections, there is a unidirectional change in the content of the 0.5–0.25 and 0.25–0.1 mm fractions along the profile.

The sediments penetrated by sections 3, 6, 8, 9 are characterised by the highest content of medium-grained sand −54–70%. Unidirectional with the prevailing fraction changes along the beach profile have sands of 0.25–0.1 mm size, multidirectional −1.0–0.5 mm. In general, the considered sediments characterise the transit areas.

Deposits characterised by a two-modal distribution character are exposed in sections 2, 7, 12. Fractions of medium-grained sand and pebbles with gravel have oppositely directed content change along the profile, unidirectional −0.5–0.25 and 1.0–0.5 mm.

If all the sediments considered above are characterised by an equally gradient, oppositely directed change in the content along the profile of the prevailing fraction (0.5–0.25 mm) and deposits of larger size, then in the vicinity of the Vagis m. (Section 4), the content of medium-grained sand and pebbles with gravel is unidirectional. Vagis (Section 4), this regularity is broken. Here a pair of 0.5–0.25 and 0.25–0.1 mm is observed. Moreover, fractions of 1.0–0.5 and 0.5–0.25 mm have the same character of distribution along the profile.

Fig. 8 Beach survey using the MDS-1 beach transect method (Piltun Lagoon embayment): (**a**) preparation of the transect for description and sampling; (**b**) orthophoto plan of the transect

The beaches of north-eastern Sakhalin were studied mainly as part of survey work in connection with coastal defence. Morphodynamic observations, core drilling, beach sections with granulometric characterisation of sediments were supplemented by GPR surveys.

As in the case of western Sakhalin and Chukotka, data on the size composition of beach sediments, structure, thickness of the active layer, and facies characteristics of the strata were obtained from beach transects (Fig. 8). In some areas, beach transects were not carried out for the full width of the beach and additional manual drilling was carried out (Figs. 9 and 10).

Methodological issues of GPR studies are discussed in the previous chapter. With regard to the beach area, interesting results obtained in the area of the coast between Lunsky and Nabil bays should be noted. The rear edge of the beach at the time of the survey had almost identical elevations along the shoreline. However, it was the results of the GPR profile analysis that revealed the alternation of high (up to 4.0 m) and low (1.5–2.0) elevations of the rear edge of gravel and pebble beach deposits overlain by fine-medium-grained sands, possibly overweathered. This allows us to assert the megafeston organisation of storm facies here.

Fig. 9 Pits for description and sampling of beach sediments: (**a**) MDS-2; (**b**) MDS-3

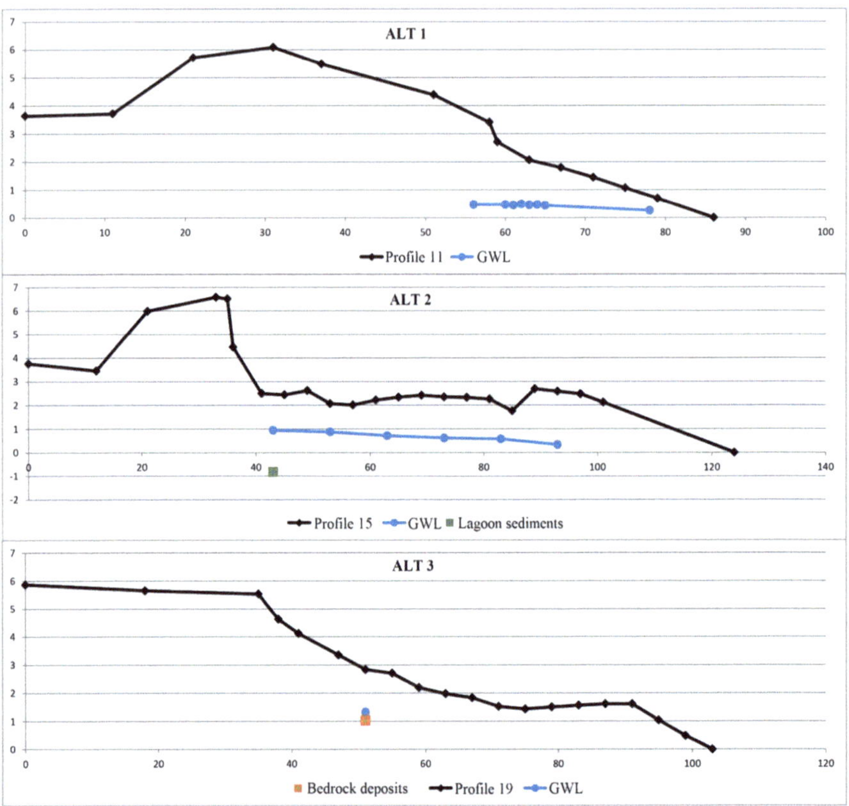

Fig. 10 Profiles of the beach studied by beach transects and boring method

Morphometric studies of the beach were carried out at all study sites of the shores of the Northern Pacific. Based on the results, morphometric maps of the beach surface and surface difference maps were constructed to determine the parameters of the resulting deformations of the upper part of the coastal slope over a certain period. The most interesting materials are presented in the following sections of the chapter.

In areas with close bedrock bedding, a combined method of beach survey was also used (Fig. 11).

In this section, we present some spectra of beach profiles indicating an uneven distribution of beach morphometric parameters along the shore (Figs. 12 and 13).

The material presented indicates that beach rear edge marks at the accumulative ledge sites are as high as 4.2 metres and at the minimum beach width sites 1.0 metre.

3 Rhythmic Patterns of Beach Material Distribution Along the Shore as a Result of the Formation of Megafestoons and Their Systems

Rhythmic or periodic structures, manifested as wave-like perturbations of the shoreline contour, are observed on many accumulative coasts of seas and large lakes. Such structures include, for example, beach festoons, megafestoons, longshore sand waves, and series of spits. While the first of them are relatively small in size, the rest of them represent rather large accumulative formations with a step along the shore of the order of hundreds of metres and an amplitude along the normal to the shore of the order of tens of metres (Bowen 1969a, b, Bowen 1984). For example, longshore sand waves appear as alternating zones of accumulation and erosion of the shore, moving in the direction of the resulting sediment flow. On the shores of Holland and Long Island, wave movement rates are measured in the hundreds of metres per year (Falqués 2005; Thevenot 1994). In contrast to sand waves, megafestons exhibit only alongshore migrations. However, even these movements cause vertical deformations of the coastal topography, measured in first metres, which have to be taken into account in engineering practice. A system of megafloods with a well-defined step of 500–600 metres is observed, for example, on Nantucket Island (Atlantic coast of the USA) (Ashton 2001). In the Tatar Strait, hybrid periodic forms several kilometres long are observed, combining systems of large-scale megafestons and obliques (Zenkovich 1962). There are also giant rhythmic structures, apparently formed over many hundreds of years. An example of giant rhythmic formations is a series of sandy spits on the northern coast of the Sea of Azov, which are about 30 km apart and extended 10–15 km into the sea.

This section deals with megafestons, which were recorded at many coastal areas of the Far East seas. On the coast of western Kamchatka, megafestons with a period from 400 to 1000 m were recorded at Oktyabrskaya and Kirovskaya Spits. On the Pacific coast of Kamchatka, the system of megafaistons with a distance between

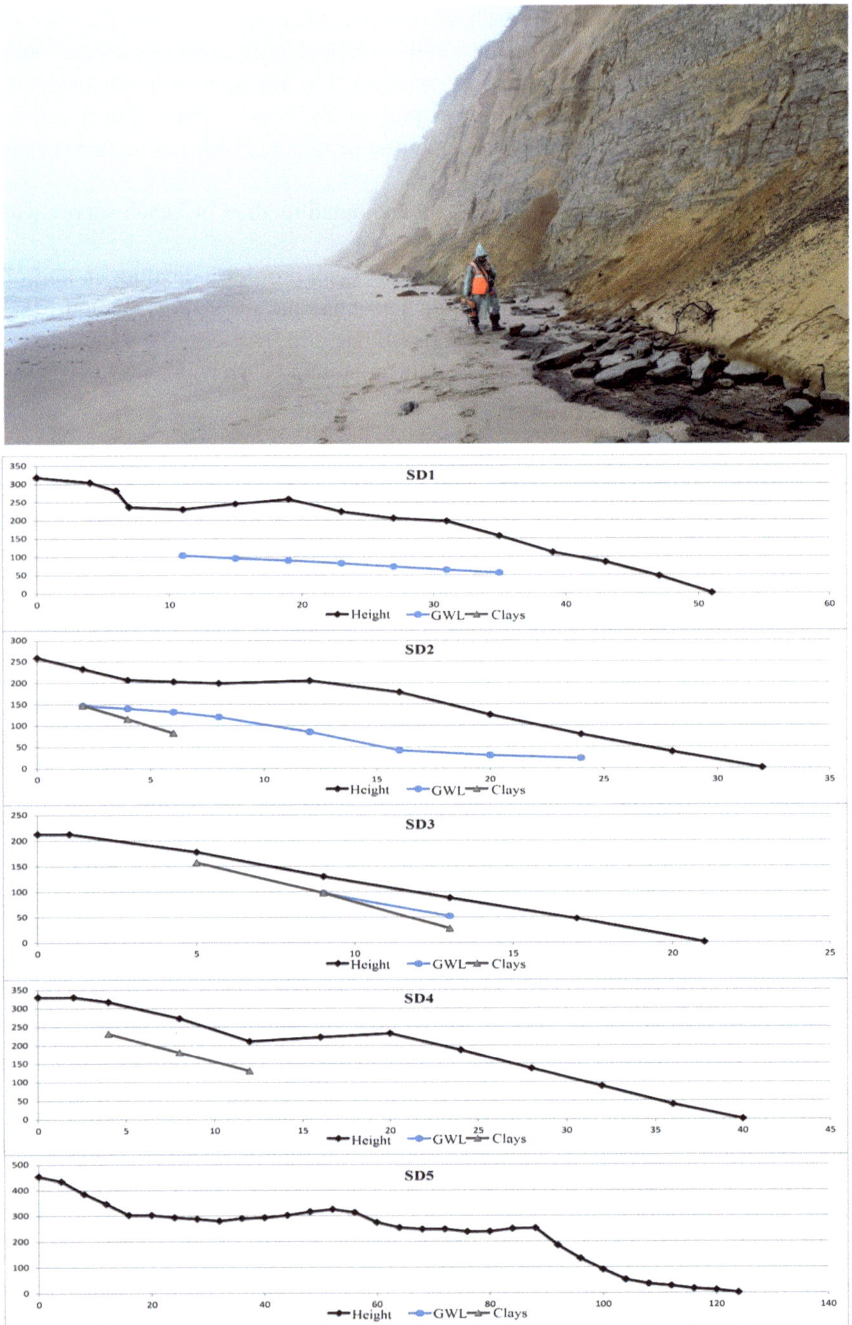

Fig. 11 Beach structure at the erosion site of a high terrace fragment composed of Neogene lithified clays

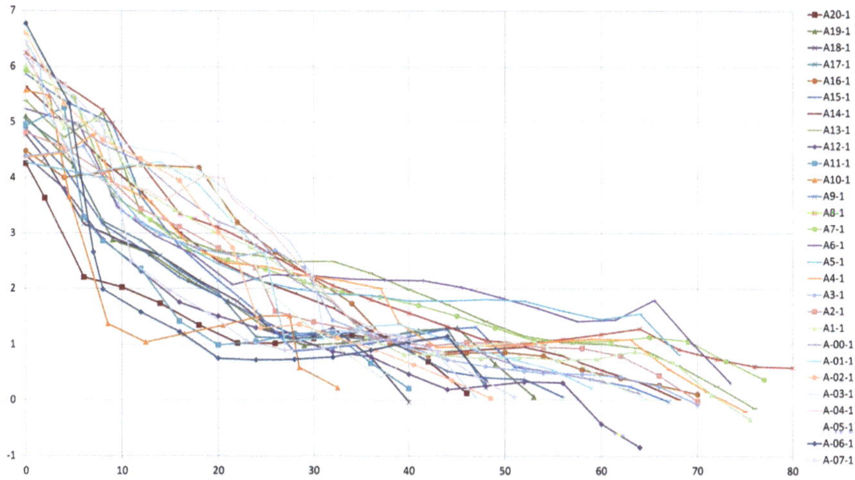

Fig. 12 Spectra of beach profiles at the Lunsko-Nabilsky section (4.4 km length)

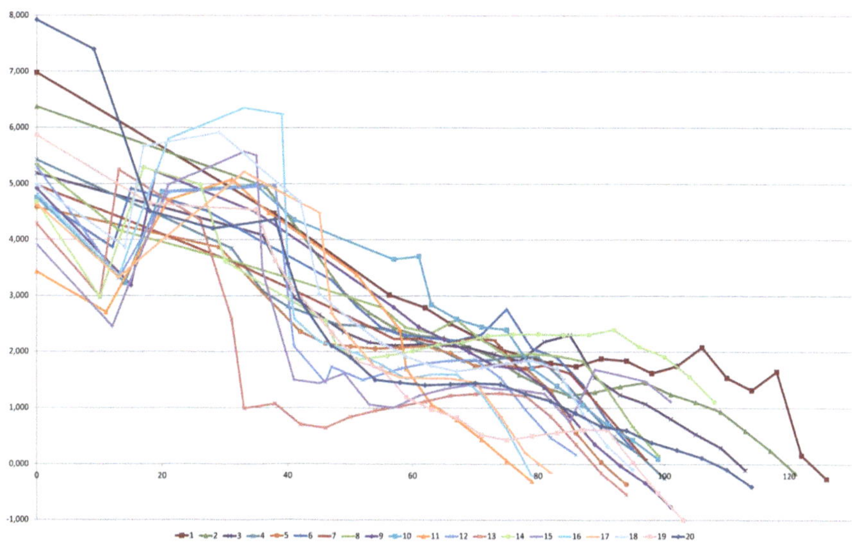

Fig. 13 Spectra of beach profiles in the Piltun section (2.0 km length)

capes of 400–900 m on the shores of Avacha Bay has become a local landmark (Fig. 14). Very large megafestons with a period of about 2 km are observed near Magadan in the northwestern part of the Taun Bay (Fig. 15).

Megafestons are widely represented on accumulative shores and shores with easily erodible low Pleistocene terraces of western Beringia. Interestingly, such structures do not occur on narrow transit sections of Holocene accumulative formations.

Fig. 14 Megafaestons of Khatyrka beach (Kamchatka)

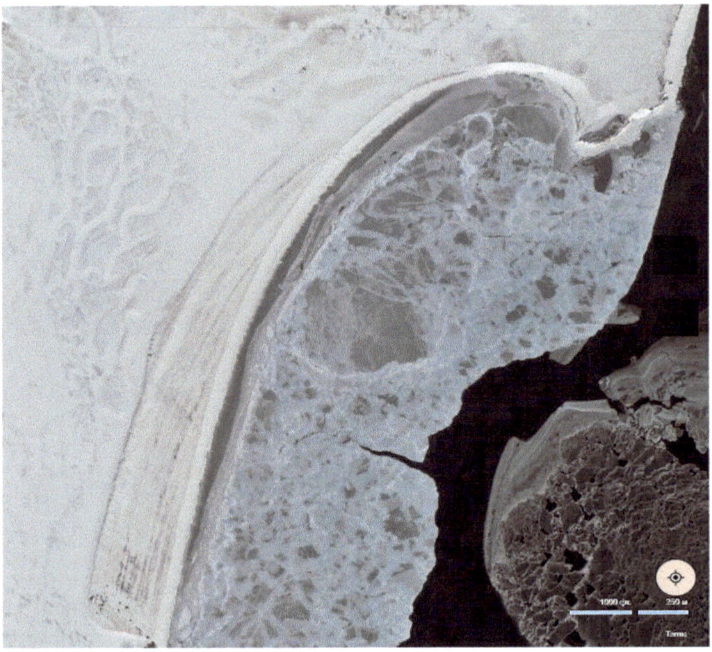

Fig. 15 Megafestons in Shelekhov Hall

Fig. 16 Megafestons near Neshkan settlement, Chukotka (fragment of photo from Bing public resource)

Fig. 17 Megafestons in the vicinity of Enurmino settlement, Chukotka (pravchukotka.ru/wp-content/uploads/2017/09/IMG_20170907_150659.jpg)

Uneven distribution of beach sediment along the shore is also characteristic of the Arctic shores of Chukotka (Figs. 16 and 17).

In the area of Neshkan settlement, the period of festoons on the marine side of the bar is approximately 2 times longer than on the lagoonal side (Fig. 16). On the Kuril Islands megafestons are observed both from the sea and ocean sides (Figs. 18, 19).

Fig. 18 Megafestons of the Okhotomorsk coast of the Kuril Islands

Fig. 19 Megafestons of the ocean coast of the Kuril Islands

And if from the Okhotomorsk side the period of these forms is 400–600 metres, from the ocean side it is already 1200–1500 metres. On Sakhalin Island, more than 40 sites where megafestons and sand waves are observed (Figs. 20, 21 and 22).

One of the coastal areas where the megafaestone system remains more or less unchanged for a long time is located in the area of Vzmorie settlement on the coast of Terpeniya Bay (Fig. 23).

The presence of megafloods leads to an uneven distribution of beach material along the shore. The width of the beach on coastal projections (headlands) reaches 45–90 m, while in concavities (bays) between headlands it does not exceed 3–8 m (at high tide).

Fig. 20 Megafaistons in the vicinity of the root north bar of Piltun Lagoon (northeastern Sakhalin)

Fig. 21 Megafestons in the vicinity of the root southern part of the Lunskoye Lagoon bar (northeastern Sakhalin)

As a result, in bays the avandune and marine terrace often experience erosion and the shore retreats, although on the whole the stock of beach material in the lithodynamic system is quite large (on average 140–150 m^3 per metre of shore length). This section presents the results of long-term observations of the dynamics of this system. The properties of megafloods and their development trends on the scale of seasons, years and decades are discussed. A model explaining the appearance and development of rhythmic forms of the coastal contour as a result of self-organisation of the relief associated with the mechanism of coastal circulation cells is proposed. The area under consideration belongs to the northern flank of the morpholithodynamic system of the coast of Terpeniya Bay. It is bounded by Cape Senyavin in the south and Cape Mulovsky in the north (Fig. 24).

The main impact on the shores is caused by SE and W swells, with a significant role played by quicksand waves. In the summer season the height of significant waves usually does not exceed 2.5 m, and in autumn increases to 4 m and more. During the passage of tropical cyclones, storm waves with a height of more than 6 m are observed in the coastal water area.

Fig. 22 Megafestons in Terpeniya Bay (south-eastern Sakhalin)

Fig. 23 Megafeston system near Vzmorie settlement in Terpeniya Bay

The maximum tide is about 1.5 m. Wind surges are insignificant (no more than 0.3 m). Ice cover (including fast ice) lasts on average from January to April.

The supply of beach-forming material to the lithodynamic system due to solid river runoff is estimated at 40 thousand tons/year. Beach sediment composition is dominated by coarse sand and gravel.

The input of beach-forming material to the lithodynamic system from solid river runoff is estimated at 40,000 tonnes/year. The beach sediment composition is dominated by coarse sand and gravel.

At present, the banks in the study area are mostly retreating. The removal of significant volumes of sand and gravel material from the coastal zone for

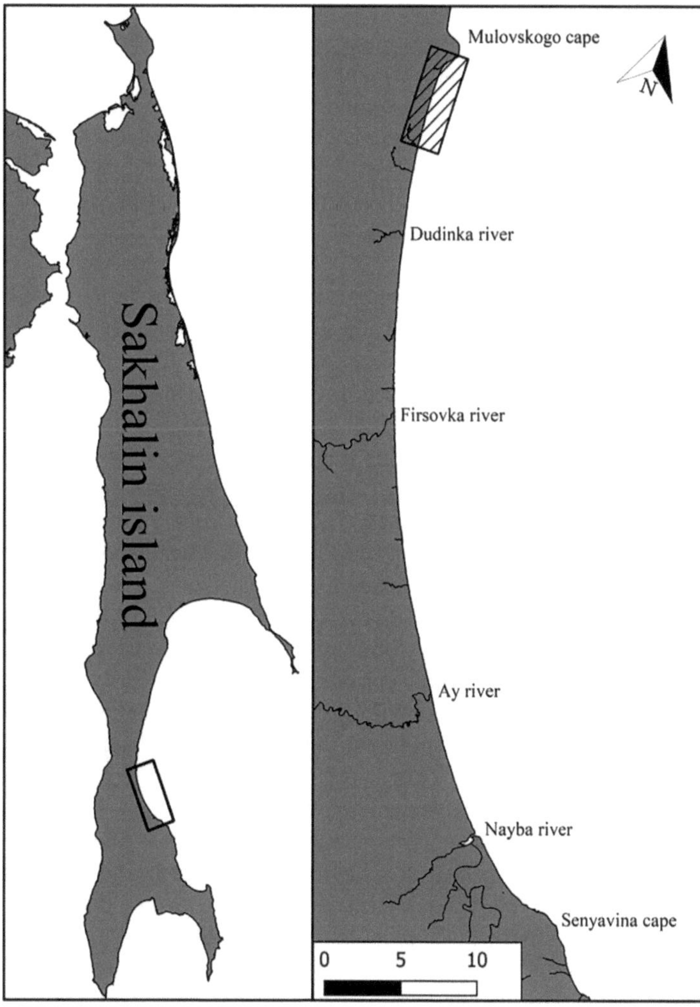

Fig. 24 Location of the detailed study area

construction purposes (about three million m^3) may have played a certain role in this process. The erosion of low marine terraces in the considered lithodynamic system has been observed since the early 1960s. One of the zones of maximum erosion is confined to the Dudino quarry, where beach sediments were mined until recently.

The Yuzhno-Sakhalinsk—Okha—Moskalvo motorway runs directly offshore in the study area (Fig. 23). A comparative morphometric analysis of coastal conditions along the route over the period 1991–2007 showed that within a 5 km stretch the beach lost between 10,000 and 50,000 m^3 of sediment annually. The losses were apparently related not only to anthropogenic influences on the sediment budget, but also to an increase in relative sea level, in particular due to tectonic submergence of the coastal land. Regional analyses allow us to take the average rate of relative level rise in the study area as 3 mm/year. Using the well-known Bruun's rule (Bruun 1986), which determines the displacement of the coastal profile, it is possible to estimate the volume of sediment that should be transported to the submarine slope as a result of sea level rise. This volume is calculated to be around 40,000 m^3/year for a 5 km stretch of beach, which is broadly in line with the actual beach loss.

The presented work implements the methodology of morphodynamic studies based on both traditional geomorphological and modern methods of obtaining and analysing geospatial information.

The arrays of aerial photo information were processed as follows. At the first stage, electronic versions of ASF swaths of aerial photographs for the period 1952–2007 were created. Then coordinate georeferencing of the raster topo map of 1:100000 scale in the free cross-platform geoinformation system QGIS (Quantum GIS) and coordinate georeferencing of all ASF strips by characteristic points was carried out. After implementation of the transformation procedures, the degree of contour coincidence of the geographical objects of the best quality strip of the ASF with the topo map drawing was checked. The most adequate variant was selected. The other ASF strips were georeferenced to it using practically the same reference points. From 1999 to 2014 the dynamics of the coastal zone was also analysed on the basis of CS interpretation in GIS environment. As a result of the processing of the obtained information, the values of coastal scarp erosion were estimated both for different stages and for the whole period.

Measurements of beach morphometric parameters and shoreline position in the megafeston system were carried out from 2006 to 2017. Observations were carried out 2–3 times a year during the ice-free period, and mainly the results of autumn surveys were used for comparative analysis. The altitudinal framework was based on the 1977 Baltic Height System. For calculations of beach volumes, a coordinate system was used, where the X-axis length measure was the distance along the motorway from the first reference point, and the Y-axis was the distance from the current reference point to the given point in the coastal profile.

The zero mark of the vertical Z-axis corresponded to the mean sea level. The point measurements thus obtained by triangulation and conversion of the resulting polygonal model into a raster grid file were converted into a terrain model with 0.5 m spacing on the Y axis and 1 m on the X axis and height on the Z axis in each cell. For all models, volumes bounded by the model surface and the horizontal surface passing through $Z = 0$ were calculated.

Inter-annual volumes of sediment change in the lithodynamic system were then obtained by subtracting the earlier model from the later one, element by element. To assess the accuracy of the results, the 2006 and 2007 model volumes were compared with beach volumes obtained for the same areas from topographic survey data. Differences were found to be within 3–4%.

Figures 25 and 26 show the results of observations characterising the beach and shoreline dynamics between 2006 and 2017.

A route survey of the site in 2005 identified 6 contrasting beach headlands. The southernmost accretion in Fig. 25 is a projecting river delta. The distance between

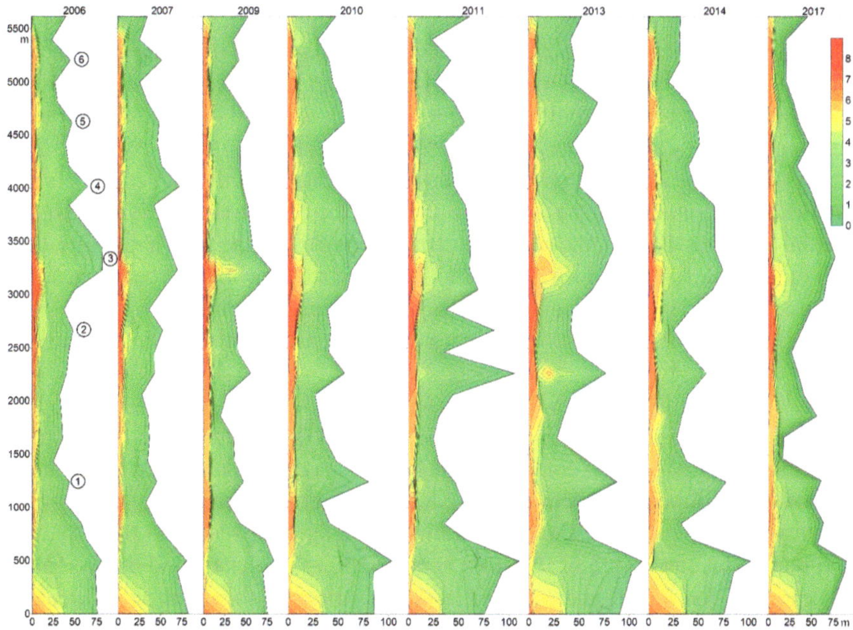

Fig. 25 Morphodynamics of the stable megafeston system (sand waves) of Terpeniya Bay for the period 2006–2017

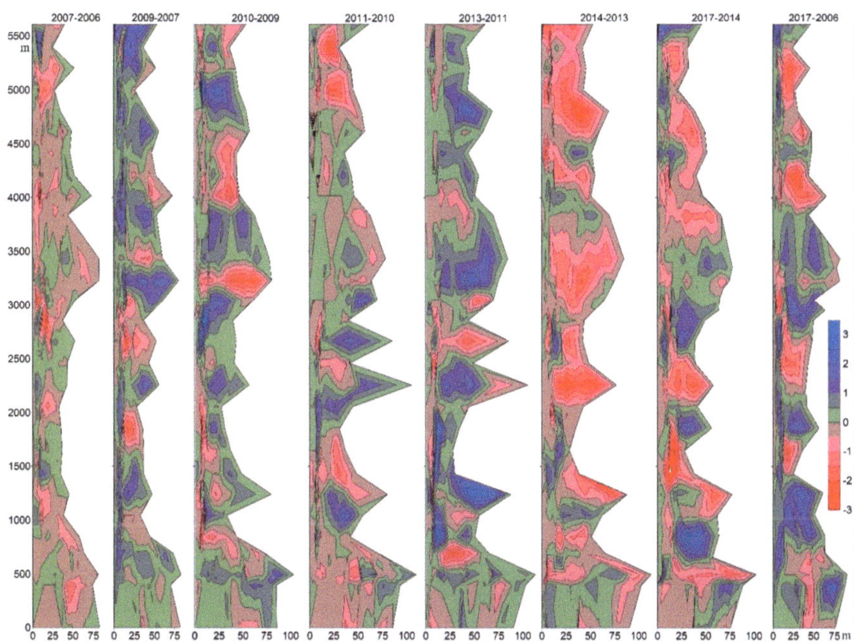

Fig. 26 Interannual deformations of the megafeston (sand waves) system of Terpeniya Bay

the first south-facing ledge and the second ledge was 1250 m, the second and third ledge 670 m, the third and fourth ledge 800 m, the fourth and fifth ledge 620 m, and the fifth and sixth ledge 540 m. By 2009, the second ledge had shifted southwards by about 400 m, and the fourth ledge merged with the third ledge to form a large accumulative ledge, which has persisted to the present day.

By the start of construction of the temporary shore protection structure between the delta ledge and the first beach cape in 2012, the distance between the first and second cape was 1070 m, the second and third cape 850 m, the third and fourth cape 800 m, and the fourth and fifth cape 730 m. The sixth cape has not been traced for the last 5 years. The sixth promontory has not been traced for the last 5 years. At the beginning of the next year, 2013, there was a shift of almost all capes to the north by 50 to 170 m.

By the end of 2013, the headlands had moved southwards by 90–200 m. This year is characterised by maximum volumes of beach sediments in the coastal morpholithosystem. While the contrast of the beach capes was maintained in 2014, the volumes of sediments in them decreased. In 2015, beach promontory № 2 shifted northward by about 50 m, and the accumulative ledge that started to form in 2014 two hundred metres south of it shifted another 100 m southward and became more contrasting. The fourth and fifth ledges have shifted to the south by 90 and 60 metres, respectively. The third ledge is practically not pronounced. By 2017, ledge №1 had shifted southwards by about 400 m, ledge № 2 by another 100 m southwards (since 2014–by 400 m).

Analyses of changes in the situation for 2006–2017 generally show an increase in the contrast of accumulative ledges in the southern half of the morpholithosystem. In the northern part there was a levelling of the coast due to the reorganisation of cells. At the same time, part of the beach material moved southwards.

The general increase in beach sediment volume that began after 2007 is primarily due to the closure of a sand quarry located 4.5 km south of the study area. However, beach volume began to decrease again from 2014 onwards and in 2017 equalled the 2009–2010 values. Changes in beach volume between 2006 and 2017 are shown in Fig. 27.

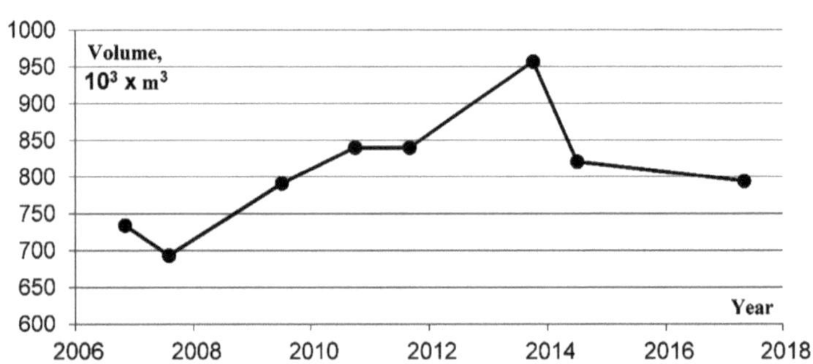

Fig. 27 Change in beach sediment volume at the megafestooned shoreline site between 2006 and 2017

Figure 28 shows the status of the megafestoon system during the summer and autumn seasons of 2010. As can be seen, in summer the shoreline contour did not change significantly and the beach slightly increased in volume (by 16,000 m^3 from the end of May to the end of August). In autumn the general picture remained the same, although in the southern part of the site the megafloor became more contrasting and the beach volume slightly decreased.

For a long time, it was believed that edge waves arising from the reflection of infragravity waves from the shore were responsible for the formation of rhythmic

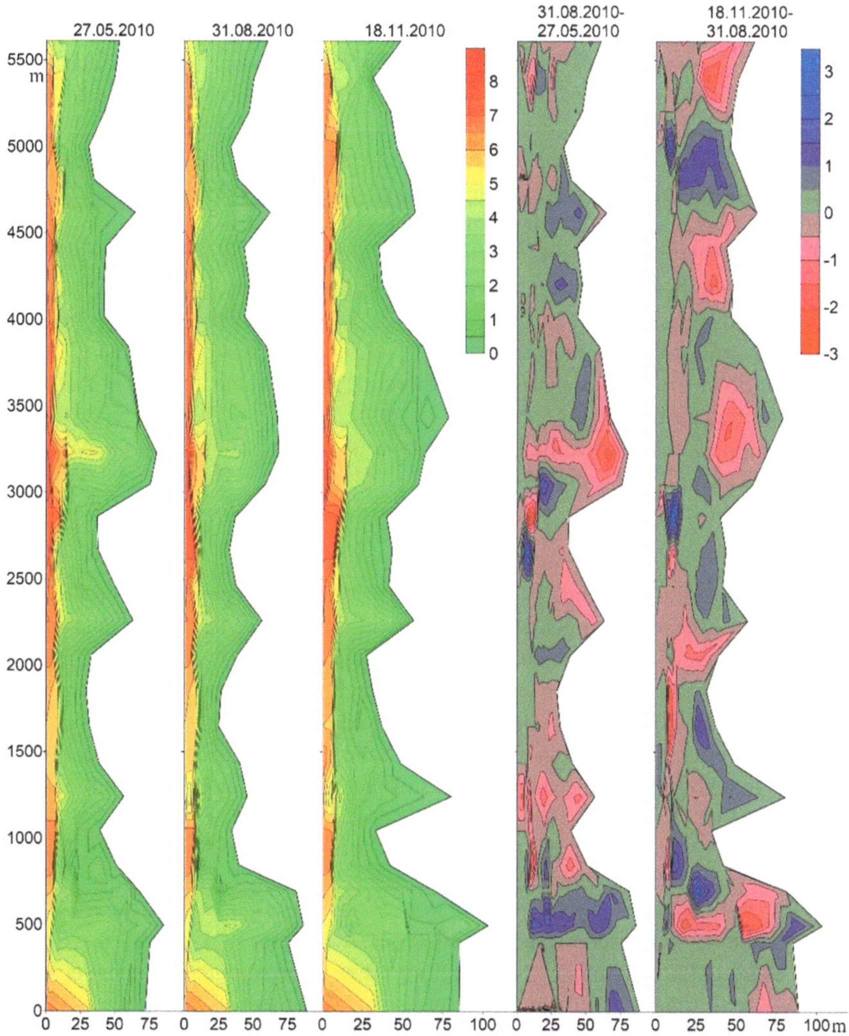

Fig. 28 Dynamics of the megafeston system in summer and autumn 2010. The left graphs represent changes in beach morphometry from late May to mid-November. The right graphs show changes in beach volume during the summer and autumn seasons

structures of the coastal contour (Bowen 1984; Holman 1982). However, despite the correlation between the size of some shapes and the probable length of edge waves, no convincing evidence in favour of this hypothesis has been obtained. There is now greater acceptance of the view of the dominant role of the mechanism of self-organisation of landforms, which is based on a positive feedback between hydrodynamics and sediment transport (Falqués 2000; Hino 1974).

The instability of an initially flat coastal contour, which is responsible for the growth of periodic disturbances, can be caused, for example, by wind waves approaching the shore at a significant angle (more than 45° relative to the normal to the shore) (Plekhanov 2014; Hino 1974). Accordingly, a wave climate characterised by a significant contribution of "large angle" waves is favourable for the development of rhythmic structures (Ashton 2005). This feature, however, is not characteristic of the studied section of the coast of Terpeniya Bay, where the dominant wave directions, in general, do not form significant angles with the coastal normal.

In our opinion, the development of rhythmic forms here is caused by other reasons. As shown below, the interaction of coastal morphology with currents in circulation cells may play a determining role in this (Leontiev 2014).

Let us consider the surf zone of wind waves propagating along the normal to the coast (along the X-axis), whose contour at the initial moment is rectilinear (parallel to the Y-axis) (Fig. 29).

During wave collapse, the mean water level rises and reaches a value ζ_0 at the shoreline. Since the conditions along the shoreline are the same, there is no horizontal water circulation at first.

Now suppose that a wave-like disturbance of small amplitude a with wave number k has occurred on the shoreline:

$$x_0 = -a \sin ky, \tag{1}$$

where x_0—coastline position, $k = 2\pi/\lambda$, λ–length (step) of the perturbation. This will immediately lead to the appearance of the level gradient $\partial\zeta/\partial y$, which determines the development of longshore currents from capes to bays.

Here, in turn, an outflow current from the shore (a prototype of a rip current) will form, and circulation cells (with trailing currents behind the collapse line) will result. The transverse velocities U in the surf zone will be distributed in accordance with the value of $-kx_0$, and the velocities of the longshore currents V will change proportionally to $\partial x_0/\partial y$, i.e.

$$U = -\hat{U}kx_0, \qquad V = \hat{V}\left(\partial x_0 / \partial y\right), \tag{2}$$

where $\hat{U}$ and $\hat{V}$—amplitude values of velocities.

Let us try to find out under what conditions the perturbation of the shore contour can grow with time. For this purpose, let us use the law of conservation of mass, which can be written in the form:

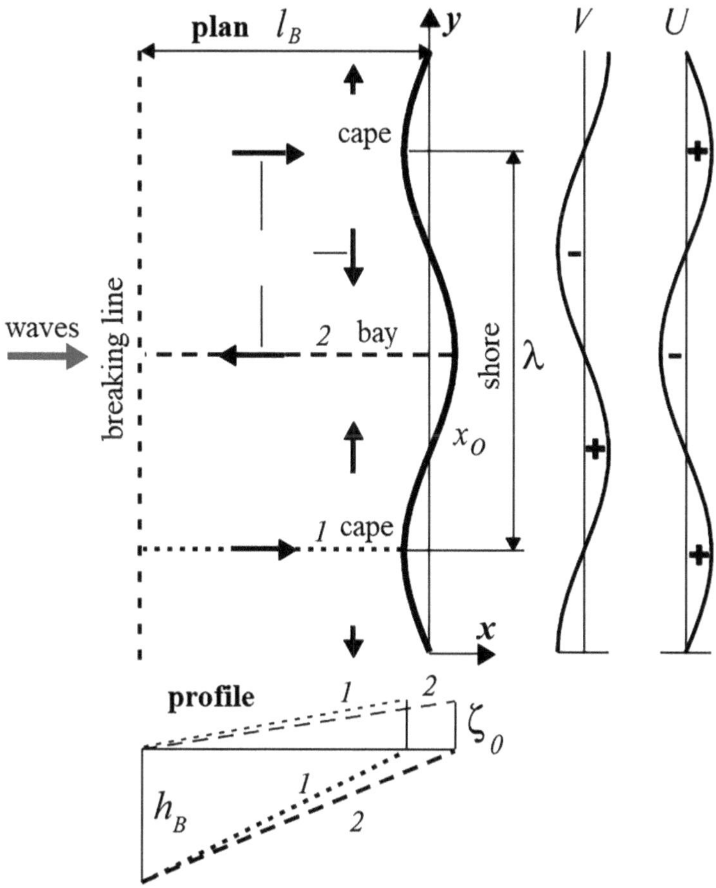

Fig. 29 Schematic diagram of currents and suspended sediment fluxes under conditions of a wavy shoreline contour. l_B—width of the surf zone, h_B—depth of collapse, x_0—position of the shoreline, ζ_0- height of the wave run-up, λ—length of the shoreline disturbance, U and V—transverse and longitudinal current velocities

$$\frac{\partial x_0}{\partial t} = \frac{\partial Q_y / \partial y - \overline{q_x}}{h_B + z_c}, \tag{3}$$

where $\partial x_0/\partial t$ – coastal drift rate, t—time, Qy—is the integral longshore sediment discharge, $\overline{q_x}$—the resultant lateral discharge through a single cross-section, h_B –the depth of collapse, z_c—the beach elevation corresponding to the height of the surge.

Equation (3) essentially expresses the fact that when longshore flow increases ($\partial Q_y/\partial y > 0$), the beach erodes and retreats ($\partial x_0/\partial t > 0$), and when sediment inflow ($\overline{q_x} > 0$), the beach accretes and extends ($\partial x_0/\partial t < 0$).

Допустим, что наносы в прибойной зоне транспортируются преимущественно во взвеси, усредненная концентрация которой равна $\overline{c}$. Тогда, приняв среднюю глубину как $\frac{1}{2}(h_B + z_c)$, определим расходы наносов:

Let us assume that the sediment in the surf zone is transported mainly in suspension, the average concentration of which is $\overline{c}$. Then, taking the mean depth as $\frac{1}{2}(h_B + z_c)$, we determine the sediment load:

$$\overline{q_x} = \frac{1}{2}\overline{c}(h_B + z_c)U, \qquad Q_y = \frac{1}{2}\overline{c}(h_B + z_c)l_B V \tag{4}$$

where l_B—the surf zone width, and the current velocities are expressed by relations (2). Substituting (4) into (3), we arrive at the diffusion equation:

$$\frac{\partial x_0}{\partial t} = \hat{v}\left(\frac{\partial^2 x_0}{\partial y^2} + \frac{\hat{U}}{\hat{V}}\frac{kx_0}{l_B}\right), \qquad \hat{v} = \frac{1}{2}\overline{c}l_B\hat{V}, \tag{5}$$

where $\hat{v}$—diffusion coefficient. After substituting (1) into (5), we obtain the equation with respect to the perturbation amplitude:

$$\frac{\partial a}{\partial t} = \left(\frac{\hat{U}}{\hat{V}}\frac{1}{kl_B} - 1\right)\hat{v}k^2 a, \tag{6}$$

the solution of which is written as $a = a^{(0)}\exp\left[\left(\frac{\hat{U}}{\hat{V}}\frac{1}{kl_B} - 1\right)\hat{v}k^2 t\right]$, (7)

where $a^{(0)}$—initial value of a at the moment t = 0.

As can be seen from (7), the perturbation amplitude will grow with time if the inequality is fulfilled

$$\frac{\hat{U}}{\hat{V}}\frac{1}{kl_B} - 1 > 0 \qquad \text{or} \qquad \frac{\lambda}{l_B} > 2\pi\frac{\hat{V}}{\hat{U}}. \tag{8}$$

Condition (8) defines the minimum disturbance length λ as a function of surf zone width and current velocities in the circulation cell.

If the transverse and longitudinal velocities are close to each other ($\hat{U}/\hat{V} \approx 1$), the length λ, according to (4.8), should be greater than $6l_B$. However, it is known from observations that outflow velocities tend to exceed the velocities of the feeding longshore currents due to the nonlinear properties of the coastal circulation (Bowen 1969a, b). In this regard, data (Brander 1999) obtained offshore Australia under quasi-equilibrium circulation cell conditions under steady moderate swell are of interest. The measured mean velocity of the rip current was 0.4–0.6 m/s, while the velocities of the feed currents were in the range 0.2–0.4 m/s. The corresponding maximum values ranged from 1.6–2.1 m/s and 1.0–1.3 m/s. In view of the above data, a typical value of the ratio $\hat{V}/\hat{U}$ can be assumed to be approximately 0.5. Then it follows from (8) that in a quasi-equilibrium system, the megafloor spacing should be close to four times the surf zone width:

$$\lambda \approx 4l_B. \tag{9}$$

Perturbations with smaller spacing have no chance for growth, as this would require too great an excess of transverse current velocities over longitudinal velocities. If the length of the disturbance is very long, the developing currents are too weak to support its growth. The observed step of rhythmic shore contour shapes does not exceed $8l_B$.

An external factor played an important role in the dynamics of the considered lithodynamic system, namely, the cessation of material removal from the coastal zone (due to quarry closure), which gave impetus to an additional cycle of its evolution with a period of about 10 years. At the beginning of the cycle the beach volume increased, then decreased and at the end of the cycle almost stabilised at a value greater than the initial one. The system thus reached a new quasi-equilibrium state.

Unrelated to the mentioned factor, megafeston fluctuations can be considered as a consequence of fluctuations in the wave environment on the scales of storms, seasons, years and decades. These fluctuations cause heterogeneity of material distribution in different parts of the system, which causes migrations and reorganisation (merging and isolation) of individual elements. For example, as the swell increases, the extent of the surf zone l_B increases rapidly, and the spacing of existing megafloods is too small to satisfy the requirement (8). Unrealistically high transverse current velocities would be required to maintain the periodic shapes as they are. Therefore, the existing system is eroded and more extended structures begin to form. However, this process requires a considerable amount of time, which is usually insufficient due to the limited duration of the stabilisation phase of a strong storm.

The creation of smaller shapes that respond to moderate wave action is faster and may be long enough to form a quasi-equilibrium system. What, after all, stops the advance of headlands and coastal retreat in bays? The deterrent seems to be wave refraction. The point is that in the head part of the rip current, where the velocity drops sharply, intensive sediment deposition takes place. The material moves in both directions along the shore, and as a result an accumulative body is formed in the form of a crescentic bar convex to the sea, which is like a mirror reflection of a festoon on the shoreline. Wave refraction over the berm reduces wave heights at headlands and increases them in bays. The longitudinal mid-level gradients driving the circulation become less contrasting. Eventually an equilibrium is established in which the volume of sediment carried by the rip current is minimised.

According to condition (5.9), in the quasi-equilibrium system of megafestons, the step should be close to the value $4l_B$. Megafestons observed in Terpeniya Bay have lengths from 500 m to 1 km. At the characteristic value $\lambda = 800$ m, the system corresponds to wave conditions with a surf zone width of about 200 m. With a characteristic bottom slope of 0.015, the corresponding collapse depth h_B should be close to 3 m. Taking into account the depth ratio, the height of collapsing waves of 1 per cent supply in the system may be about 2.4 m, the height of significant waves is $H_s = 1.6$ m, and the average height is $\bar{H} = 1.0$ m. Thus, megafloods in Terpeniya Bay are mainly supported by moderate waves, which are characterised by significant recurrence throughout the year.

The question remains why, given sufficient material reserves, the system of rhythmic forms is stable in some parts of the coastline, while in others it is not. Probably, an important role here is played by local peculiarities of underwater relief and wave refraction, which influence the properties of the wave field in the coastal zone.

The megafloor system in Terpeniya Bay exists in conditions of general coastal retreat caused both by natural processes and, until recently, by anthropogenic impact. Long-term observations reveal significant fluctuations in the elements of the system on the scales of seasons, years and decades. Nevertheless, its basic morphological features have persisted for more than 60 years. The characteristic step (or length) of megafloods is about 800 m, and the range of shoreline fluctuations reaches 40–50 m.

The analysis shows that the emergence of the megafeston system is closely related to the formation of circulation cells, due to which a small initial perturbation of the coastal contour can increase with time. A positive feedback between topography and hydrodynamics occurs when a certain relationship between the spacing of rhythmic shapes λ, the size of the surf zone l_B and the velocities of coastal currents is fulfilled. In other words, the length of the initial disturbance must fall within the desired range of values.

The growth of megafestoons continues until an equilibrium is reached between their length and the parameters of the current wave. The system step is close to 4 times the surf zone width ($\lambda \approx 4l_B$). On this basis, it can be assumed that the megafeston system in Terpeniya Bay is supported mainly by moderate waves with the height of significant waves H_s of about 1.6 m.

4 Conclusion

Long-term studies of beach structure and morphometry for practical design purposes on the shores of Sakhalin Island have led to the conclusion that a fully wave-absorbing profile is formed on sand and gravel beaches at sediment volumes of 70–90 m^3/m. There is no sediment deficit on most of the accretionary shores of Sakhalin Island, but scour is observed even at average beach sediment volumes within lithodynamic cells of 140–150 m^3 per 1 metre of shore length. In our opinion, scouring of such shores is related to the uneven distribution of beach material along the shore as a result of the formation of megafairstones and their systems. This is illustrated by the example of three sites with observed shore scarp failure, where, given the uniform distribution of beach material within the lithodynamic cell, the possibility of scouring of marine terraces in Sakhalin Island conditions is virtually ruled out.

References

Dolotov, Yu.S. Dynamic conditions of coastal-marine relief formation and sedimentation [Text] / Yu.S. Dolotov. - Moscow: Nauka, 1989. - 269 p.

Belberov, E.K. Results of studies of the coastal zone dynamics under the programme "Kamchia" [Text] / E.K. Results of studies of the coastal zone dynamics under the programme "Kamchia" [Text] / E. K. Belberov, S. M. Antsyferov. - M.: IOAN SSSR, 1984. - 58 p.

Shuisky, Y.D. Problems of research into the balance of sediment in the coastal zone of the seas [Text] / Y.D. Shuisky. - L.: Gidrometeoizdat, 1985. - 240 p.

Leontiev, I. O. Dynamics of the surf zone [Text] / I. O. Leontiev. - M., 1989. - 184 p.

Kosyan, R.D. Hydrogenic movements of sediments in the coastal zone of the sea [Text] / R. D. Kosyan, N.V. Pykhov - M.: Nauka, 1991. - 280 p.

Aibulatov, N.A. Dynamics of solid matter in the shelf zone [Text] / N. A. Aibulatov. - L., Gidrometeoizdat, 1990. - 271 p.

Horn, D.P. Beach groundwater dynamics / D.P. Horn // Geomorphology. – 2002. – P. 48(1–3). – P.121–145. – (DOI:https://doi.org/10.1016/s0169-555x(02)00178-2)

Leontiev, K. Fundamentals of geomorphology of sea shores [Text] / K. Leontiev - M.: Izd-vo MSU, 1961. - 412 c.

Díez, J. Characterization of the dry beach profile: A morphological approach / J. Díez, V. Cánovas, A. Uriarte, R. Medina // Journal of Coastal Research. – 2017. – 33(6). – P. 1292–1304. – (https://doi.org/10.2112/JCOASTRES-D-16-00179.1)

Osborne, P.D. Sand re-suspension events in a high energy infragravity swash zone / P.D. Osborne, G.A. Rooker // J. Coast. Res. – 1999. – 14. – P. 74–85.

Hughes, M.G. Flow velocity and sediment transport in the swash zone on a steep beach / M. G. Hughes, G. Masselink, R. W. Brander // Mar. Geol. – 1997. – V136. – P. 91–103.

Beach, R.A. Infragravity driven suspended sediment transport in the swash, inner and outer-surf zone. / R.A. Beach, R. W. Sternberg // Proc. Coastal Sediments '91. – ASCE, 1991. – P. 114–126.

Elfrink, B. Hydrodynamics and sediment transport in the swash zone: a review and perspectives / B. Elfrink, T. Baldock // Coastal Engineering. – 2002. – T. 44. – №. 3–4. – P. 149–167. – (doi: https://doi.org/10.1016/s0378-3839(02)00032-7)

Erikson, L. Prediction of swash motion and run-up including the effects of swash interaction / L. Erikson, M. Larson, H. Hanson // Coastal Engineering. – 2004. – T. 52. – №. 3. – P. 285–302.

Dolotov, Yu. S. Sediment differentiation and layering of coastal sediments [Text] / Yu. S. Dolotov, R. B. Zharomskis, V. I. Kirlis. - Moscow: Nauka, 1982. - 184 p.

Leontiev, K. K. Leontiev, L. G. Nikiforov, G. A. Safyanov - Moscow: Izd vo MSU, 1974. - 336 p.

Wright, L. D. Morphodynamic variability of surf zones and beaches; a synthesis / L. D. Wright, A. D. Short // Marine Geology. – 56 (1984). – P. 93–116. – (https://doi.org/10.1016/0025-3227(84)90008-2)

Short, A. Waves – dominated beachs / A. Short // In: SHORT, A. (Ed.). Handbook of Beach and Shoreface Morphodynamics. – Chichester, John Wiley & Sons, 1999. – P. 173–191.

Castelle, B. 2007. Double bar beach dynamics on the high-energy meso-macrotidal French Aquitanian Coast: a review / B. Castelle, P. Bonneton, H. Dupui, N. Sénéchal // Marine geology. – 245(1–4) – P. 141–159.

Masselink, G. The effect of tide range on beach morphodynamics and morphology: a conceptual beach model / G. Masselink and A. D. Short // Journal of coastal research. – 1993. – P. 785–800.

Short, A. D. Single and multi-bar beach change models / A. D. Short, T. Aagaard // Journal of Coastal Research. – 1993. – P. 141–157.

Aagaard, T. Sediment transport on dissipative, intermediate and reflective beaches / T. Aagaard, B. Greenwood, M. Hughes // Earth-Science Reviews. – 2013. – №124. – P. 32–50. – (https://doi.org/10.1016/j.earscirev.2013.04.002)

Hughes, M. G. Spectral signatures for swash on reflective, intermediate and dissipative beaches / M. G. Hughes, T. Aagaard, T. E. Baldock, H. E. Power // Marine Geology. – 2014. – V354. – P. 88–97. – (https://doi.org/10.1016/j.margeo.2014.04.015)

De Vries, S. Aeolian sediment transport on a beach with a varying sediment supply / S. De Vries, S. M. Arens, M. A. De Schipper, R. Ranasinghe // Aeolian Research. – 2014. – T. 14. – P. 235–244.

De Vries, S. Dune behavior and aeolian transport on decadal timescales / S. De Vries, H. N. Southgate, W. Kanning, R. Ranasinghe // Coastal engineering. – 2012. – T. 67. – P. 41–53. – (https://doi.org/10.1016/j.coastaleng.2012.04.002)

Dean, R. G. Cross-shore Sediment Transport Processes / R. G. Dean // Advances in Coastal and Ocean Engineering. – 1994. – vol. 1. – World Scientific. – P. 159–220.

Díez, J. Spatial and Temporal Variability of Dissipative Dry Beach Profiles in the Pacific Northwest, USA / J. Díez et al. // Journal of Coastal Research. – 2016. – T. 34. – №. 3.

Hesp, P. A., 1986. Surf zone, beach and foredune interactions on the Australian south east coast / P. A. Hesp // Psuty, N. (ed.), Dune Beach Interaction. Journal of Coastal Research, Special Issue. – No. 3. – P. 15–25

Short, A. D. Three dimensional beach stage model / A. D. Short // Journal of Geology. – 1979. – V.87. – P.553–571.

Afanas'ev, V. V. Morphology and dynamics of the shores and bottom of the Nevelsky Strait - the area of the permanent railway crossing design [Text] / V. V. Afanas'ev, E. I. Ignatov, S. V. Chistov. - Smolensk: Izd-vo MSU, 2006. - 128 p.

Afanas'ev, V. V. V. Geomorphological structure and morphodynamics of the north-western coast of Sakhalin Island / V. V. Afanas'ev // Dissertation. Candidate of Geographical Sciences. Moscow State University, 1996. – 185p.

Komar, P. P. Beach processes and sedimentation / P. P. Komar // New York: Prentice-Hall, 1975. – 430 p.

Bowen, A. J. Rip currents. Part 2. Laboratory and field observations / A. J. Bowen, D. L. Inman // J. of Geophys. Res. – 1969a. – V. 74. – № 23. – P. 5479- 5490.

Bowen, A. J. Waves, long waves and nearshore topography / A. J. Bowen, D. A. Huntley // Marine Geol. – 1984. – V. 60. – P. 1–13.

Falqués, A. Wave driven alongshore sediment transport and stability of the Dutch coastline / A. Falqués // Coastal Eng. – 2005. – V. 53. – P. 243–254.

Thevenot, M. M. Longshore sand waves at Southampton Beach, New York: observation and numerical simulation of their movement / M. M. Thevenot, N. C. Kraus // Marine Geology. – 1994. – V. 125. – P. 249–269.

Ashton, A. D. Formation of coastline features by large-scale instabilities induced by high-angle waves / A. D. Ashton, A. B. Murray, O. Arnault // Nature. – 2001. – V. 414. – P. 296–300.

Zenkovich, V. P. Fundamentals of the doctrine of the development of sea coasts [Text] / V. P. Zenkovich. – M.: Publishing House of the USSR Academy of Sciences, 1962. – 710 p.

Bruun, P. The Bruun rule of erosion by sea-level rise: a discussion on large-scale two- and three-dimensional usages / P. Bruun // J. of Coastal Res. 1986. V. 4. № 4. P. 627–646.

Holman, R. A. Bars, bumps and holes: models for the generation of complex beach topography / R. A. Holman, A. J. Bowen // J. of Geophys. Res. – 1982. – V. 87. – № C1. – P. 457–466.

Falqués, A. A mechanism for the generation of wave-driven rhythmic patterns in the surf zone / A. Falqués, G. Coco, D. A. Huntley // J. of Geophys. Res. – 2000. – V. 104. – № C10. – P. 24071–24087

Hino, M. Theory on formation of rip current and cuspidal coast / M. Hino // 14th Int. Conf. Coast. Eng. ASCE. – 1974. – P. 901–919.

Plekhanov, F. A. Spatial and temporal variability of wave characteristics off the south-eastern coast of Sakhalin Island according to instrumental measurements / F. A. Plekhanov // Scientific Notes of Sakhalin State University. - 2014. - №. 1. - P. 35-36.
Ashton, A. D. High-angle wave instability and emergent shoreline shapes: 2. Wave climate analysis and comparisons to nature / A. D. Ashton, A. B. Murray // J. of Geophys. Res. – 2005. – V. 111. – F04012. – (Doi:10. 1029 / 2005JF000423)
Leontiev, I. O.O. Morphodynamic processes in the coastal zone of the sea / I. o. o. o. Leontiev. - LAP LAMBERT Academic Publishing. Saarbrücken, Deutschland (Germany), 2014. - 251 c.
Bowen, A. J. Rip currents. Part 2. Laboratory and field observations / A. J. Bowen, D. L. Inman // J. of Geophys. Res. – 1969b. – V. 74. – № 23. – P. 5479- 5490.
Brander, R. W. Field observations on the morphodynamic evolution of a low-energy rip current system / R. W. Brander // Marine Geol. – 1999. – V. 157. – P. 199–217.

Conclusion

Research has shown that three morpholithodynamic stages are generally recognized in the development of the coasts of subarctic and moderately cold seas during the autumn-winter period. Initially, the beach freezes, flattened aufeis forms, and coastal escarpment erosion intensifies due to the weakening of the beach's wave-damping properties during the cold period. Several weeks after the onset of freezing of beach sediments, stable marginal aufeis forms, the steepness of the coastal profile increases, and the depths in the near-shore zone correspondingly increase, leading to increased splash height and a coarser grain size distribution of sediments. With the formation of ice cover, tidal currents beneath fast ice continue to reshape the underwater topography. Deformations on bay shores can reach 1.5–2.0 m. Based on the developed concept of coastal zone dynamics during the transition of average daily temperatures to subzero values, periods of maximum risk of coastal erosion have been identified for these seas. It is concluded that the estimated period of hazardous impacts on the coastal cliff corresponds to the data of field observations and confirms the proposed concept of coastal zone development after the transition of average daily temperatures to subzero values.

The main patterns of the spatial structure of coastal erosion determine the rhythmicity and cyclicity of erosion processes. Analysis of the graphs of the spectral density of erosion intensity, determined on the basis of processing in a GIS environment of aerial photography (flights of 1952, 1974) and space photography (2014 and 2015) allowed us to establish erosion rhythms with a wavelength along the coastline of 1.3, 2.0, 4.0, 5.6, 8.5, 11.6, 25.6, 64 km. High-order rhythmicity of erosion processes, in our case $\approx$ 30–60 km, on leveled shores is associated with stable zones of divergence of the resulting sediment flows, characterized by a reduced volume of beach deposits. The erosion rhythm of 1.3 km corresponds to the dimension of large megafestoon beach structures.

To form a fully wave-damping profile on the studied sand and gravel beaches of Sakhalin, 70–90 m^3 of sediment per linear meter of coastline is sufficient. However,

T. Chaplina (ed.), *Coastal Geosystems of Cold Seas Under Conditions of Climatic and Anthropogenic Changes*, Earth and Environmental Sciences Library, https://doi.org/10.1007/978-3-032-12247-6

coastal erosion is observed even with average beach sediment volumes within lithodynamic cells of 140–150 m^3/linear meter. The erosion of such shores is caused by the uneven distribution of beach material along the coast resulting from the formation of megafestoon systems. Stable megafestoon systems are maintained primarily by moderate wave action, with significant wave heights of approximately 1.6 m. Despite significant fluctuations on seasonal, annual, and decadal scales, the main morphological features of stable systems have been preserved for at least 60 years.

It was found that the largest Holocene sedimentary formations are confined to areas with the highest sedimentation rates throughout the Neogene-Quaternary period. Changes in the morphometric parameters of sedimentary formations over the period 1952–2012 indicate sediment accumulation in areas of inherited accumulation, amidst a general trend of increasing coastal erosion observed since the 1940s.

At this stage of coastal development in the Far Eastern seas, sediment budget parameters determine the erosion of coastal escarpments in unconsolidated Neogene-Pleistocene sediments and the reorganization of Holocene depositional forms. The greatest morpholithodynamic changes in the coastal zone of the lagoonal shores of the North Pacific occur in strait systems. Analysis of data, specifically on coastal reorganization of northeastern Sakhalin, showed that, on average, approximately 13,000 m^3 of sediment per year enters the sediment convergence zone near the Nyivo Lagoon Strait from the north over a long-term period. Approximately 106,000 m^3/year, of which approximately 65,000–70,000 m^3 is open-sea sediment (siltstone-pelite fractions), enters from the south. In the Nyivo Lagoon strait zone, according to the analysis of coastal re-formation maps, approximately 23 thousand m^3/year of beach-forming fraction sediments accumulate, which practically equalizes the incoming and outgoing parts of the sediments of the constructed balance model of the lithodynamic system of northeastern Sakhalin.

High-precision digital photogrammetric multi-temporal models (SFM Photogrammetry) of the abrasion-denudation coastlines of the North Pacific seas were constructed, allowing us to determine the rates and nature of failure of bedrock of weak and moderate stability. High-precision data on the failure rates of a coseismically uplifted bench surface revealed that the failure of the subhorizontal bench surface during the adaptation stage of the subaqueous relief type to subaerial conditions occurs at rates approximately an order of magnitude lower than the failure rates of subvertical coastal bluffs in rocks of weak and moderate stability. The role of wave action in bench development is primarily reduced to the mobilization and removal of debris resulting from frost weathering at subzero air temperatures during the open sea period. High-precision orthophotomaps and 3D models of a high coastal bluff composed of volcanogenic-sedimentary rocks allowed us to identify the failure mechanisms of cliffs with heterogeneous strength and determine their quantitative parameters.

Only recently, when analyzing geospatial information on Arctic coastal dynamics, has it become clear that minimal displacements are observed on spits and bars, while maximum erosion rates are observed on the shores of plains formed by lake drainage. The area of barrier islands on the western coast of the Beaufort Sea increased by 45% between 1940 and 2010. Interestingly, barrier islands on Russia's delta coasts are also stable or growing. We established sediment accumulation in areas of inherited accumulation with increasing coastal erosion observed since the 1940s when analyzing the dynamics of Sakhalin's sediment formations. This means that bars and spits, which maintain their connection to their recharge areas, do not degrade but merely rebuild themselves in both subarctic and Arctic coastal conditions. The most vulnerable to coastal reshaping are Shishmaref, Kivalina, and Unalakleet, located at the distal ends of the spits of the lagoonal strait system. Accordingly, coastal protection solutions should be based on an analysis of strait migration, the shape and morphometry of the spits' distal ends, modifications to the tidal prism, and the potential for changes in spits' growth direction.

MIX
Papier aus verantwortungsvollen Quellen
Paper from responsible sources
FSC® C105338

If you have any concerns about our products,
you can contact us on
ProductSafety@springernature.com

In case Publisher is established outside the EU,
the EU authorized representative is:
Springer Nature Customer Service Center GmbH
Europaplatz 3, 69115 Heidelberg, Germany

Printed by Libri Plureos GmbH
in Hamburg, Germany